Connecticut FARMS AND FARMERS MARKETS

Tours, Trails *and* Attractions

ERIC D. LEHMAN
AND AMY NAWROCKI

Globe Pequot
Essex, Connecticut

Globe Pequot

An imprint of Globe Pequot, the trade division of The Rowman & Littlefield Publishing Group, Inc.
4501 Forbes Blvd., Ste. 200
Lanham, MD 20706
www.rowman.com

Distributed by NATIONAL BOOK NETWORK

British Library Cataloguing in Publication Information Available

Library of Congress Cataloging-in-Publication Data Available

ISBN 978-1-4930-6585-1 (paper : alk. paper)

ISBN 978-1-4930-6586-8 (electronic)

∞™ The paper used in this publication meets the minimum requirements of American National Standard for Information Sciences—Permanence of Paper for Printed Library Materials, ANSI/NISO Z39.48-1992.

CONTENTS

ACKNOWLEDGMENTS

We would like to thank David and Trena Lehman, Dana Jackson, David K. Leff, Elizabeth Hindinger, Jamie Jones, Lee Edwards, Francois Steichen, and our editor Amy Lyons. We are also grateful to Patti Popp of Sport Hill Farm; Samantha Sabia of Sabia Tree Farm; John Halfinger of Halfinger Farm; Lawrence Kalbfeld and Barbara Puffer from the Guilford Agricultural Society; Amanda Freund of Freund's Farm Market and Bakery; and Lydia Casparie of Arethusa Farm for additional photographs. Most of all, we would like to thank the farmers and farms featured in this guide, for continuing the life-giving tradition of agriculture into the 21st century. Their work continues to inspire us.

INTRODUCTION

Slowing down off the highway exit, you find yourself on a pleasant country road, leaving the suburbs and heading into the forested hinterlands. Suddenly, the green tunnel opens and cows loll their tongues over the stone walls toward you. A wooden farm stand with a handmade sign beckons from the roadside, and you pull into a sandy turnout and park. The farmer greets you and offers a fresh strawberry. You pop it into your mouth, the sweet, bold flavor taking you back to childhood. You fill a basket with peppers, pears, and crisp green onions. Continuing down the quiet lanes, you find another farm selling alpaca wool hats, and another boasting fresh eggs and milk. At a crossroads a market bustles with life on a smooth village green. Others have found this marvelous agricultural world, too, just off the beaten track, right here in beautiful Connecticut.

Split by the river that gives its name to the state, our landscape was carved by glaciers, which also left innumerable rocks on and in the earth, often called "New England potatoes," that would be the bane of farmers." The soft peaks of the western highlands, the alluvial plain of the coastal regions, and the rocky dells of the eastern uplands all have their own unique climates and microclimates. Over the centuries, this landscape has been transformed many times, from fire-scorched parkland, to lush green farmland, to smokestack factory metropolis, to forested residential suburbs.

The region's indigenous peoples farmed the land long before European settlers arrived, planting sunflowers and Jerusalem artichokes, cultivating oyster and clam beds, and establishing the Three Sisters system of maize, beans, and squash to create a balanced nutritional diet. European colonists learned some of these techniques, but also tried their own crops, such as apples, beets, and potatoes. Connecticut pork became famous in the 1700s and Connecticut peaches became a surprise success in the 1800s. With the rise of industry, agriculture declined as the primary mover of the state's economy, with standouts like chickens and tobacco remaining popular in the 20th century. Today, as a densely populated state with only 10 percent rural population, our farms still number about 6,000, and in recent years we have experienced renewed interest in specialty smallholdings by enterprising young farmers and locally sourced food by consumers.

We may visit farms and markets because we want to support the local economy. We may go because we want to improve the health of the environment. Or we may go because we are interested in improving our personal health. Perhaps all these reasons, and more. Even as our world becomes more interconnected and we can order food at the touch of a button, knowing where our food comes from and how it is treated has become more important to all of us. This may seem like looking to the past, but maybe we are also looking forward. We need relationships with our neighbors, and we need good food. We need these things like we need oxygen, and when we don't have them, we begin to decline, as individuals and as a society. Everyone knows this, though often in our globalized, online world we don't want to admit it.

You know it, at least, because you have opened this book and are thinking about where to go to pick some apples, sample some wine, or buy a gallon of freshly made ice cream. Thinking about how to improve your relationship with the land, with your agricultural neighbors, with your home. In that way, this is a guide not just to Connecticut farms and markets: it is a guide to a better life.

Ten Ways to Support Connecticut Farms and Farmers

Shop Local, Buy Direct: Shopping locally makes sense for health, economic, and environmental reasons. When purchasing food grown at the farm, your money goes directly to the people who planted, cultivated, and harvested it. Large farm stands also sell crafts, canned items, packaged food, and crops from other small producers. Support a network of local distribution rather than large agri-corporations. When you go to grocery stores, choose locally sourced vegetables, fruit, dairy and meat products. Pledge to Go

Local through BuyCTgrown (buyctgrown.com) and spend 10 percent of your food and gardening budget on locally grown and made items.

Sign Up for a CSA: Participating in a Community Supported Agriculture program lets you pre-pay for the season's harvest and gives you a weekly portion to take home. CSAs give farms guaranteed income necessary to keep doing what they do best. CSA members get the freshest pick, and you can expect to see your selections change week by week as the seasons develop. Some farms share only what they grow on-site, but with others you can expect flowers, honey, and baked goods sourced from other farms. Full shares or

half shares are usually available; pickup days vary so contact your farm directly.

Pick Your Own: There's no better way to connect with the land than to get out into the field and gather the earth's bounty yourself. Whether you're planning to pickle cukes, fill jelly jars, gather Halloween pumpkins, or chop your own Christmas tree, Connecticut's farms have you covered season to season. Grab the kids, your hat and gloves and hit the field. If you plan accordingly, and pick up a recreational shellfish permit, you next CT Grown picnic will be impressive!

Taste Wine, Buy Wine: Don't let *Wine Spectator* or the *New York Times* tell you what you should drink. Visit Connecticut vineyards and tasting rooms and sample first. Keep an open mind and prepare to be surprised. Then make a purchase. Bring local wine to your next gathering. Buying directly helps the vineyard grow, adapt to nature's fickle seasons, and plan the next vintage. This goes for breweries and distilleries that source locally, too. A growing network of hop (cthopgrowers.org) and grain farmers in the state means CT brewed can be CT grown.

Choose Local Fish: Climate change, pollution, and overconsumption affects the food chain no matter where we live. And supporting local aquaculture means thinking and acting with purpose about what goes on your plate. Connecticut's shores and rivers yield many Connecticut specialties: bluefish, shad, and trout, not to mention oysters, clams, and lobster. Grab your pole and put a line in on the banks of the Farmington. Your fishing license fees support conservation and wildlife management efforts. Watch an expert bone the delicate (but bony) shad at an early summer festival, then enjoy the delicate whitefish yourself. Shop at fish markets and support restaurants that serve local fish. Ask the grocery store fishmonger to carry less popular (but very tasty) fish that are sourced responsibly and locally.

Vote with your Wallet: Purchasing food and goods from a farm stand supports local agriculture directly. But there are other ways to do more with your money. Support nonprofit organizations directly with donations. Shop at sites that stream donations into local organizations, such as Amazon Smile, which puts 0.5 percent of eligible purchases toward organizations like the Connecticut Agricultural Education Foundation or the Connecticut Farmland Trust. It may not seem like a lot, but it will remind you of core values when you need to shop online.

Support Farm-to-School Programs and Agricultural Education: Feed the future and help our children eat better and know more about where food comes from. Encourage local school boards to invest in programs that give school kids access to better nutrition and farms a source of income. Better still, encourage schools to grow gardens and diversify their cafeteria menus with student-grown vegetables, herbs, and flowers. Plant pride and community learning.

Rent a Goat, Walk a Llama, Adopt an Animal: Support animal farms and rescue sanctuaries and the animals they care for by spending some time with the herd. Connect with them for all your goat-scaping needs and reduce the need for pesticides on poison ivy. Sign up for a llama hike or goat yoga and exercise your body and that of your new friend. Visit and support places like Locket's Meadow Farm in Bethany, which cares for animals rescued from slaughter, abuse, and neglect.

Reduce Waste: Amy's father had a saying: take all you want; eat all you take. In our family of seven, he wanted to avoid leftovers that sat in the fridge, spoiled, and ended up in the trash. As an adult, Amy likes to think that what he really meant was to be respectful of what our mother had done to prepare the food; take into consideration that others aren't as lucky, and act wisely about what's on the plate. A bruise on your apple doesn't mean the apple is no longer edible. The expiration date on your gallon of milk is not an invitation to throw half of it out the next day. Think creatively about what will become of the food you put in your basket. If you're not going to use it all, take less. If you can, invest in a composting barrel. Put trash in proper receptacles. Recycle what you can. Donate leftovers to food banks.

Spread the Word: Small enterprises rely on word-of-mouth and community connections to grow and flourish. In the age of social media, that means everyone can play a part as an agritourism ambassador. Hashtag your photos and give a shout-out to local vendors. "Like" and follow farms and organizations on social media sites. Sign up for emails alerts and newsletters. Share good news, reviews, and event updates.

A Guide to Seasonal Fruits and Vegetables

No matter how often we frequent farmers markets, each of us would benefit from becoming a little more in tune with nature's growing cycles. In Connecticut, changing seasons are part of our identity, so we should take better advantage of seasonal fruits and vegetables, looking for fresh spinach in spring, sweet corn in summer, or squash in autumn. We might even get inspired to pickle and preserve our summer bounty for the long winter.

Here is a harvest guide to help plan your shopping:

April: arugula, asparagus, cabbage, chard, fiddleheads, herbs, kale, spinach
May: arugula, asparagus, beets, broccoli, broccoli rabe, cabbage, cauliflower, chard, collards, fava beans, fiddleheads, herbs, kale, lettuce, new potatoes, parsley, parsnips, pea greens, peas, radishes, ramps, rhubarb, scallions, spinach
June: arugula, asparagus, beets, bok choy, broccoli, broccoli rabe, cabbage, carrots, cauliflower, chard, cherries, collards, corn, cucumbers, fava beans, herbs, kale, kohlrabi, lettuce, oregano, pea greens, peas, peppers, radishes, raspberries, rhubarb, scallions, snap beans, spinach, strawberries, summer squash, turnips, zucchini blossoms

July: apples, apricots, beans, beets, blackberries, blueberries, bok choy, broccoli rabe, cantaloupe, carrots, chard, cherries, collards, corn, cucumbers, eggplant, garlic, green beans, herbs, kale, kohlrabi, lettuce, lima beans, melons, nectarines, okra, onions, oregano, parsley, peaches, peas, peppers, plums, potatoes, radishes, raspberries, rhubarb, scallions, snap beans, spinach, strawberries,

summer squash, tomatoes, turnips, winter squash, zucchini, zucchini blossoms

August: apples, apricots, arugula, beans, beets, blackberries, blueberries, bok choy, broccoli rabe, cantaloupe, carrots, celery, chard, collards, corn, cucumbers, currants, eggplant, garlic, grapes, green beans, herbs, kale, leeks, lettuce, lima beans, melons, nectarines, okra, onions, oregano, parsley, parsnips, peaches, pears, peas, peppers, plums, potatoes, radishes, raspberries, rhubarb, rutabagas, scallions, snap beans, summer squash, sweet potatoes, tomatoes, turnips, watermelons, winter squash, zucchini

September: apples, apricots, arugula, beans, beets, bok choy, broccoli, broccoli rabe, cabbage, cantaloupe, carrots, cauliflower, celery, celery root, chard, collards, cranberries, cucumbers, eggplant, garlic, grapes, green beans, herbs, kale, kohlrabi, leeks, lettuce, lima beans, melons, nectarines, okra, onions, oregano, parsley, parsnips, peaches, pears, peas, peppers, plums, potatoes, pumpkins, radicchio, radishes, raspberries, rutabagas, scallions, shelling beans, spinach, summer squash, sweet potatoes, tomatoes, turnips, watermelons, winter squash, zucchini

October: apples, beans, beets, broccoli, broccoli rabe, Brussels sprouts, cabbage, cantaloupe, carrots, cauliflower, celery, celery root, chard, collards, corn, cranberries, cucumbers, eggplant, fennel, garlic, grapes, herbs, kale, kohlrabi, leeks, lettuce, lima beans, melons, okra, onions, oregano, parsley, parsnips, pears, peas, peppers, potatoes, pumpkins, radicchio, radishes, raspberries, rutabagas, scallions, shelling beans, spinach, turnips, watermelons, winter squash

November: apples, beets, broccoli, broccoli rabe, Brussels sprouts, cabbage, carrots, cauliflower, chard, collards, cranberries, fennel, garlic, kale, leeks, onions, parsley, parsnips, pears, peppers, potatoes, pumpkins, rutabagas, shelling beans, spinach, turnips, winter squash

December: apples, beets, cauliflower, cranberries, garlic, leeks, parsnips, pears, potatoes, turnips, winter squash

Year-round: cider, dairy, honey, maple products, mushrooms, protein (beef, bison, chicken, eggs, fish, lamb, pork, shellfish, turkey, veal)

Why Farm Education Is Important for Kids of All Ages

Have you ever milked a cow? Harvested corn? Sheared a sheep or wrangled a goat? For thousands of years, the vast majority of human beings did all these things. But during the past two centuries, more and more people have lost touch with these activities, and thus with the land itself. Is it any wonder that we care so little about our environmental impact? We don't see it every day, like a farmer does. Today, farm and ranch families make up less than 2 percent of America's population, with less than 10 percent involved in food production. One US farm feeds an average of 166 people.

For the rest of us, who may have experience with a home garden at most, experiencing agricultural education gets us back to the earth and helps us realize what it means that this small percentage of humans is responsible for almost all our food. It is a profound shift, and we have not yet seen how it will affect our species, much less the planet itself. Westport chef and Wholesome Wave founder Michel Nischan says that "those who remain on the land," from multi-generational farmers to young college grads to veterans returning from war to become stewards, are all "heroes."

Indeed, they are. But that is a big change from a time, not so long ago, when almost everyone was involved in the production of some or all of their own food. Farm education means developing a closer relationship with that food again, returning to something both practical and magical. Along with the opportunities outlined in this book, there are countless others waiting for you, from real, honest-to-goodness farm work to a pleasant walk with llamas through a forest. Take them. When a child first learns where a carrot comes from, pulling it out of the rich, black soil of a naturally composted field, she appreciates something greater than a fact. She learns, perhaps, to understand the very mysteries of life.

SOUTHWEST CONNECTICUT

As one of the most densely populated areas of the country—and the richest—the southwest corner of Connecticut might not seem at first to be a place for farming. However, it is still dotted with family farms and small holdings and for hundreds of years has been one of America's centers of aquaculture. From the "gold coast" to the charming hills of the interior, this corner of the state boasts an incredibly diverse range of agritourism opportunities, hidden behind the highway walls and city spires. The wealth of Fairfield and New Haven counties has allowed an increased interest in organic and natural foods, with places like the Westport Farmers Market and Sport Hill Farm modeling sustainability for the rest of us. You can spend a weekend at the Guilford Fair or a whole year returning again and again to a place like Shelton's Jones Family Farm for different agricultural experiences.

"We are fortunate to live in southwest Connecticut," says Jamie Jones, the sixth generation of his family to run his 500-acre farm. "We've been here for over 150 years and we're in a good place. People come here to experience a product that was grown and produced here—from the land."

Creameries and Dairy Farms

Ferris Acres Creamery

144 Sugar St., Newtown; 203-426-8803; ferrisacrescreamery.com

Founded in 1864 by William Ferris, this dairy farm is still owned by the family today. The red ice-cream shop in front of the silo and barn in Newtown is a local and regional favorite, offering dozens of flavors of ice cream, milk shakes, floats, sundaes, and sorbet, including vegan options. They'll also custom-craft ice cream cakes and pies for you if you call ahead to order or have the ice-cream gurus cater this year's birthday party, wedding, or summer fling. Try the "Campfire" ice cream (vanilla with fudge swirls, graham cracker pieces, and mini marshmallows) in a waffle bowl. That's right, not a waffle cone, a waffle bowl. Yum! Come back for Route 302 Chocolate Moo and watch happy cows grazing on the hill as you wait patiently in the long lines. This working family farm has better hours than most summer ice-cream shops, open Monday through Saturday 11 a.m. to 8:30 p.m. (with a special express window open at 10:30). On Sunday they close at 5 p.m. Plenty of time to indulge and sample special fall and winter flavors.

Field View Farm

707 Derby Turnpike, Orange; 203-795-0571; facebook.com/pages/FieldViewFarm/516339938387192

Field View Farm is one of the oldest farms in Connecticut, and one to visit and support as an example of endurance and dedication. Established in 1639 by Thomas Hine and his family, the farm has been the heart and soul of twelve generations of dairy farmers, producing milk and homemade ice cream. All of this despite a devastating fire in 1996 and the ongoing obstacles that face all family farms, large and small. Stop by to pet the cows, and take home some eggs, milk (including raw milk), or other seasonal items. Chat with the farmers who continue to tend to the farm they love and help them tell the story of this land for the next hundred years. Open Monday through Friday 7 a.m. to 5 p.m. and Saturday and Sunday 9 a.m. to 5 p.m.

Old Bishop Farm

500 South Meriden Rd., Cheshire; 203-439-0783; oldbishopfarms.com

Founded in the 1700s, Old Bishop embodies the best characteristics of "old fashioned." These traditions bear fruit (literally) today, with attention to quality and customer service. The creamery produces Super Premium ice cream, made on the premises, with an indulgent 16 percent cream. Small batches, big taste. Whether you like the classics (real berries in your scoop of strawberry) or prefer a little dazzle—like Black Raspberry Crunch or Honey Lavender Blueberry (seasonal)—you won't be disappointed. Inside the Country Store, you can browse gifts, kitchen and garden ware, then fill your basket with garlic scapes, cukes, greens, and a pint of strawberries—or pick your own. Delicious pies, pastries, donuts, and more are baked on-site. Take a seat on the Country Store's wood tables and grab an extra spoon to share your Donut Sundae. Or stroll the flower garden with your cone. Open Wednesday through Sunday 8 a.m. to 8 p.m.

Plasko's Farm

670 Daniels Farm Rd., Trumbull; 203-268-2716; plaskofarm.com.

Creamery, cafe, corn maze, pumpkins—Plasko's has it all, including a rich history and the significance of being the last remaining farm in Trumbull. Plasko's has kept focus on traditions begun when sixteen-year-old Martin Plasko arrived from Czecholsovakia and saved enough money to purchase land in 1925 that the next generation would build upon. Since then, the family has added more land, diversified crops, and developed a nursery, country store, and cafe. Most recently John Jr. and his wife, Pauline, have opened a creamery, adding ultra-premium homemade ice cream to the cafe's enticing menu of specialty drinks and freshly baked pastries, pies, scones, and donuts. Ask for a scoop of Cinnabar in your cold brew or go all-in with Ice Cream Affogato—two shots of espresso over your favorite scoop, plus whipped cream.

The indoor space is rustic, spacious enough for a small group, and cozy enough for a quiet lunch in front of the fire. But outside you get a fuller sense of seasonal family fun, like a corn maze that features a newly designed route through the four-and-a-half-acre field every year. There will be pumpkins to decorate and hayrides to delight and ice-cream making to watch. Come back in late November for wreaths and garlands and a cup of hot cider. The creamery is open every day 12 to 9 p.m., while the cafe is open daily 6:30 a.m. to 9 p.m. During the fall, the corn maze is open on weekends 11 a.m. to 6 p.m. No reservations are needed.

Rich Farm Ice Cream

691 Oxford Rd., Oxford; 203-881-1040; richfarmicecream.com

Nutmeggers love their ice cream (who doesn't?) and the votes are in from Hartford *Courant* readers: Rich's takes the top spot. There is good reason to scream for flavors like Mint Marshmallow Brownie and Caramel Toasted Almond. Indeed, the Department of Agriculture thinks so too, recognizing Rich Farm for both beauty and high quality. The Rich family has been operating dairy farms in the state for five generations, the tradition starting on Ajello's Farm in Seymour. From that early start, the family moved the farm to Oxford, and that's where great-grandsons Don and Rich took over. You can't miss the towering silos and green roof of the historic barn, but Rich Farm ice cream is also available at multiple locations, so you won't miss the chance to find your perfect scoop. The shop in Oxford is open Monday through Saturday 11:30 a.m. to 9 p.m., and Sunday 11 a.m. to 8 p.m. Visit the handsome "barnfront" shop on Station Road in Brookfield (open 12 to 9 p.m.), or head up to the Bristol shop, open daily 12 to 8 p.m.

Shaggy Coos Farm

53 Center Rd., Easton; 484-788-1769; shaggycoos.farm

Run by Brittany Conover, Easton's Shaggy Coos Farm raises livestock humanely, without artificial aids, feeding them antibiotic-free and hormone-free grain, hay, and grass. The cows roam the pastures, and their fresh milk is creamy and delicious, pasteurized on-site and sold at their farm stand. The stand-out product from this microdairy is their cocoa-rich chocolate milk. "People love to know the cows their milk comes from," Conover says. "They want to know that they are being taken care of." This female-run farm also sells chicken eggs, half-portion pigs, and even holiday turkeys. The farm stand uses the honor system. Fill your bag with fresh veggies (from nearby Silverman's Farm) then head into the small rustic barn to pick up your milk from the fridge and beef, bacon, and wings from the freezer. Enter your purchases, swipe your credit card, and sign your receipt. Farmer Britt's Bulgarian yogurt sells out quickly, so get there early. Join the herd—open daily 7 a.m. to 8 p.m.

Farms and Farm Stands

Ambler Farm

257 Hurlbutt St., Wilton; 203-834-1143; amblerfarm.org

Ambler Farm took shape before the turn of the 19th century, when Josiah Raymond purchased land, built his home, raised his family, and began the legacy that continues today. By 1799 Josiah, his son, and grandson, had established 300 acres and built the homes that would stay in the family for the next 200 years. Great granddaughter Hannah Raymond married Charles Ambler in 1800, inherited the farm, and had children of their own who continued, generation after generation, to work the farm until 1988. But the town of Wilton could not let the farm fade into memory, purchasing the establishment in 1999. The Friends of Ambler Farm recognized the need to honor these agrarian roots and foster the agricultural legacy begun by the Raymond and Ambler families. Today the nonprofit and a crew of volunteers work the organic farm, using sustainable practices that enhance biodiversity of the soil and yield beautiful produce, herbs, and flowers. Visit the gardens, animals, and historic buildings all year, dawn to dusk. Events and programs include summer camp, maple syruping demonstrations, yoga at the Red Barn, and much more. The farm stand operates June through October at two locations: at the farm Saturday 9 a.m. to 2 p.m., and at the Wilton Farmers Market every Wednesday from 12 to 5 p.m., located at the Wilton Historical Society (224 Danbury Road).

Bishop's Orchard

1355 Boston Post Rd., Guilford; 203-453-2338; bishopsorchards.com

Bishop's has been in operation for more than 140 years, providing the finest farm products to residents up and down the shoreline for six generations. Open year-round and conveniently located on Route 1 in Guilford, the market is vast. Inside you'll find the bakery, filled with fresh pies, breads, donuts, and other goodies made on-site. There is fresh produce from the farm that changes with the seasons, prepared foods (Bishop's own sauces, pickles, and more), and a wide selection of local meats. You can taste their wines,

especially their apple wines, on premises: Celebration 1871 tends toward the dry side, while Stone House White features a sweeter sip. In addition to wine, you'll find the pick of the harvest in Bishop's Apple Cider, classic sweet, or fermented hard cider. They also partner with Killingworth Cranberry Bog (killingworthcranberries.com), Connecticut's only bog, and with Van Wilgen Garden Center.

Stop by April through October to pick up hanging baskets, seedlings, and potted plants. The market is open Monday through Saturday 8 a.m. to 7 p.m. and Sunday 9 a.m. to 6 p.m. The winery opens a little later and closes a little earlier, and tastings are available Saturday and Sunday 12 to 5 p.m. We haven't even mentioned the creamery yet. The ice-cream window is open Sunday through Thursday 11 a.m. to 9 p.m. and Friday and Saturday 11 a.m. to 9 p.m. And one more thing! Pick-your-own seasons run June to October, for berries, peaches and pears, apples and pumpkins. Check the seasonal schedule or call the hotline for crop updates (203-458-PICK). It's easy to see why Bishop's Orchard continues to contribute so much to Connecticut.

Clover Nook Farm

50 Fairwood Rd., Bethany; 203-393-2929; clovernookfarm.com

Like so many working farms, Clover Nook is all about family and working hard to sustain a thriving farm. The farm's history dates back

to the 1600s when the French family raised livestock on the land. In 1765, David French and his wife, Hannah Lines, built their home on the Bethany land, and after David participated in the Revolutionary War, he came back to raise a family and farm. Over the decades, members of the family have added to the farm's production with hay, sweet corn, cattle, and more. Today the Demander family is the eighth generation to preserve the legacy on Clover Nook, which has been recognized as a Connecticut Century Farm. With the help of a Farm Reinvestment Grant, the industrious family added a retail farm store in 2016. There you'll find a full selection of fruits, vegetables, meat, and specialty items like heirloom tomato pasta sauce. Sign up for CSA and get a line of credit for the store, choosing items like lamb, pork, and beef, all pasture raised without hormones. The store also features products from other farms including local honey and flavored oils. Open Monday through Friday 10 a.m. to 6 p.m. and Saturday and Sunday 9 a.m. to 6 p.m.

DeFrancesco Farm

336 Forest Rd., Northford; 203-484-2028; defrancescofarm.com

DeFrancesco Farm has been growing quality vegetables since 1907, and they're still growing, now in the fifth generation. The farm's 120 acres are maintained with sustainable farming and soil management practices, using rotational crops, ground cover, and conservation irrigation. In the field, they grow cabbages to cantaloupe and much more. They also have 10 acres of greenhouses with annuals, perennials, seasonal favorites like hardy mums for fall and evergreen wreaths for winter, and an impressive selection of statuary and garden decor. They even sell soil, compost, and fertilizer for all your gardening needs. Sign up for a CSA with full and medium share options and pick up your share weekly from the end of June to the end of September. The farm stand is fully stocked with DeFrancesco veggies and preserves, along with locally produced maple syrup and honey, pastas and sauces, and dairy products from area farms like Hastings and Arethusa. Your basket will already be full, but don't miss out on their selections of pies, cookies, pastries, and donuts, all made fresh on-site. Visit Monday through Friday 9 a.m. to 6 p.m.,

Saturday 9 a.m. to 5 p.m., and Sunday 9 a.m. to 4 p.m. Hours change seasonally so check for updates, or order online for pickup.

Gazy Brothers Farm

391 Chestnut Tree Hill Rd., Oxford; 203-723-8885; gazybrothersfarm.net

The Gazy family farms 80 acres in Oxford, growing vegetables, herbs, plants, and flowers, as well as hay on neighboring land. Ed and his wife, Alexis, and brothers Pete and Tony (and their children) keep alive the practices started by Grandma and Grampa Gazy in 1918. They use Integrated Pest Management to grow beans, peas, and salad greens; peppers, potatoes, and purple asparagus; tubers, tomatoes, cabbages, cucumbers, and much more, including flowers, herbs, and fruit. Fill your basket at the farm stand or sign up for a CSA share. The farm stand is open daily from 8 a.m. to 6 p.m. mid-May through mid-October. You can also find them on different days of the week at many Fairfield County farmers markets.

Guardians Farm

99 Bates Rock Rd., Southbury; 203-906-6372; guardiansfarm.com

Guardians Farm is aptly named. As animal lovers and law enforcement specialists, David and Tamara named their farm after their "guardians of the night" K-9 partners, Chase and Anouke. Since 2014, they have been guardians of 200 acres, carrying on the tradition that was started at this Southbury farm in 1850. The authentic red barn and concrete silo house the herd of Nigerian Dwarf and LaMancha goats and look over the pasture of Holsteins and Jersey cows. Anouke, who has retired from her police duties, guards the happy animals. You'll spot Annie the mule, Kamie and Legacy (two of the goats), Sokanon the cow, and chickens pecking happily. Then check out the market for a full array of goat milk soaps, lotions, and other local products, like coffee from Great Minds Coffee Roasters in Oakville, and syrup from Maple Craft in Sandy Hook. You can also pick up Guardian's products from other local specialty shops. Visit the farm Saturday and Sunday 11 a.m. to 3 p.m.

Henny Penny Farm

673 Ridgebury Rd., Ridgefield; 203-610-9903; hennypennyfarmct.com

Henny Penny Farm began as part of the Nehemiah Keeler Tavern and Barn, helping to restore natural habits, teaching about sustainable living, and offering visitors a chance to see what a New England farm might have been like in colonial times. The family farm opened to the public in 2015, broadening their mission to maintain nature's balance through responsible farming. With a focus on fiber, they expanded their flock of Romney sheep, raised on 10 acres of town conservation land. Members of their flock—like Zeus, Chubby, Annabel, Fern, and Amelia—consistently win accolades for best fleece and best-in-show from organizations and festivals across the Northeast. The store was added in 2017, and you can purchase pasture-raised meat, eggs, dairy, and specialty products from Henny Penny and area farms. Wool products include hats, socks, handwoven Berger rugs, and artisan blankets. Knitters can choose from a variety of yarn types and colors, all from the flock. Visit the farm to see them graze, or sign up for private events, group tours, or garden club–sponsored workshops. The farm and store are open Tuesday 10 a.m. to 1 p.m. and Saturday 10 a.m. to 2 p.m. You can also find them at the New Canaan Farmers Market, Saturday, 10 a.m. to 2 p.m.

The Hickories

136 Lounsbury Rd., Ridgefield; 203-894-1851; thehickories.org

Whether from their fields, their flock, or their flower gardens, everything from The Hickories is sourced with care with a solid commitment to sustainability, conservation agriculture, and a healthy landscape for the region. In addition to certified organic fruits, vegetables, and "field to vase" flowers, the farm extends their dedication to the humane care of animals that produce meat and eggs as well as wool fibers from Finn-merino ewes. Their "ewe-nique" wool products include hand-woven scarves, shawls, socks, and hats, and are available at the store or online. Their dedication to the agricultural community and to us, the consumers, is enhanced through workshops and restoration projects, like CT NOFA's Ecotype Project. This supports development of pollinator pathways, and the farm grows and collects seeds from their plants to help the growth of native plants across the region. They have also worked to help local fish preservation by introducing trout and other native fish to their pond. All their farm products, as well as dairy and pantry items, are available at the farm stand, which is open Wednesday through Friday 2 to 6 p.m., and Saturday and Sunday 10 a.m. to 4 p.m. Tour the farm and help further support The Hickories by signing up for a CSA share.

Hindinger Farm

835 Dunbar Hill Rd., Hamden; 203-288-0700; hindingersfarm.com

Since 1893, Hindinger Farm has been growing farm fresh produce on its fields and orchards in Hamden. William and Rose Hindinger started it as a dairy farm, and their children shifted to apples, pears, peaches, and vegetables after son George returned from World War II. Today, the family's 120 acres are cultivated and managed by third-generation Anne and her children Liz and George and their families. The popular stand is stocked with Hindinger's farm fresh produce—from summer's leafy greens, onions, and radishes, to peaches in summer and apples in fall. Fully stocked shelves offer specialties like Vidalia onion salad dressing, four bean salad, and preserved vanilla peaches. But that's not all: dairy from the fridge;

local ice cream and pies; Connecticut honey and maple syrup; goat milk soap; kitchenware, gifts, and decorations; and toys and books for the kids. Speaking of kids, don't forget to visit the goat pen to say hello to Gigi, Peacherine, Milly Maddie, and Maddy, Nigerian dwarf goats who have the best view of the far-off New Haven skyline. Sign up for the mailing list and check Facebook for other events, like the June Strawberry Festival, the October Fall Festival, and food trucks with live music throughout the summer. CSAs are available for a 22-week season of freshly picked and selected produce, or as a farm credit system that allows participants to purchase credits ($100s each) for buying items from the farm store. Open April through November from Tuesday through Friday 9 a.m. to 6 p.m. and Saturday and Sunday 9 a.m. to 5 p.m.

Hollandia Nurseries and Farm

103 Old Hawleyville Rd., Bethel; 203-743-0267; hollandianurseries.com

Hollandia Nursery has been growing since 1964, starting in a garage and eventually moving to Haven Farm to bloom into one of Connecticut's premier garden centers. A visit to their extraordinary 20 acres of gardens, gazebos, and shops will inspire your creativity, and their experts will guide your selection of annuals, perennials, topiary, soil care, and growing products at the Gift and Garden Center (95 Stony Hill Road). Don't miss the spring flower show with tours, lectures, and sale prices. Then pick up your seeds, vegetables, herbs, or any of the million-plus annuals bursting with color, and perennials for next year's beauty. Their fall festival features antique tractors, hayrides, pumpkins, and colorful mums, and soon it will be time for winter garlands and mistletoe. The nursery is open daily 8 a.m. to 5 p.m., and the garden center is open daily 9 a.m. to 6 p.m.

Laurel Glen Farm

247 Waverly Rd., Shelton; 203-305-9179; laurelglenfarm.com

At Laurel Glen, you're a member of the family connected through the generations of the Rogowski family, farming their Shelton farm since the 1900s. Randy and Victoria took over the family farm in 2013 and today grow over fifty vegetable crops utilizing the Integrated Pest Management System, an environmentally sound, economic approach to pest management. Their produce reflects their commitment to sustainability and growing delicious, nutritious, and healthy food for the community. They bring their bounty to neighbors, friends, and the larger community through subscriptions, at the farm store, and at the Shelton, Monroe, and Trumbull farmers markets. The vegetable subscription offers 32-, 20-, and 13-week options of fresh, seasonal shares. At the store, you can also find meat, eggs, cheese, crackers, pastas, coffee, and more from local vendors May through mid-December, Monday through Friday 10:30 a.m. to 6 p.m., and Saturday 9 a.m. to 4 p.m.

Massaro Community Farm

41 Ford Ridge, Woodbridge; 203-736-8618; massarofarm.org

Massaro is truly a community farm. Certified organic, this nonprofit farm is a collaboration between the town of Woodbridge and the community, run by a board of directors. Their credo—"Keep Farming. Feed People. Build Community."—is apparent in everything the organization supports. Their 12 acres of cultivated fields produce vegetables, flowers, and honey that support CSA shares, farmers markets, and farm-to-table restaurants as well as many on-site events (like Dinners at the Farm). Educational programs and activities include "Kids Dig Farms" after-school events, an explorer's series for ages 4–11, beekeeping workshops (conducted by the CT Beekeeper's Association), and the nature trail. The Walk the Farm Trail lets visitors explore 45 acres of woods and wetlands and is open year-round (no dogs please). Visit their table at the Wooster Square and Edgewood Park farmers markets, or one of the farm-to-table restaurants (like Olmo, Zinc, and Heirloom) that source their produce. The Barn is open Monday through Friday 8:30 a.m. to 5 p.m.

Mitchell Farm

51 Purchase Brook Rd., Southbury; 203-264-1588; mitchellfarmct.com

Mitchell Farm has been in operation since 1759, and the family is still going strong, now with the eighth and ninth generations working 500 acres. On the farm, they harvest hay, manage firewood, grow fresh produce in season, and sell farm-fresh eggs. At the farm stand (and at area farmers markets) you can find freshly picked sugar snap peas, green beans, and romaine lettuce. Another few weeks and you can find bite-size plums, juicy tomatoes, golden zucchini, and yellow watermelon. Don't miss their specialty—corn in yellow, white, and butter-and-sugar varieties. At the farm stand, you can pick up veggies as well as eggs, maple syrup, honey, and a bundle of kindling. Call for firewood delivery. The self-serve farm stand is open year-round, daily from 9 a.m. to 4 p.m.

Muddy Roots Farm

175 Northford Rd., Wallingford; 203-213-6001; muddyrootsfarm.com

Krista Marra and Chris Wellington are the heart and soul of Muddy Roots Farm. Both have strong agricultural roots: Krista grew up on land that belonged to the Cooke dairy farm, which operated into the 1990s on the land that she farms today. Chris grew up in Salem, milking cows after high school before joining the Marines. Together they work to show that a small plot of land that is nurtured sustainably with organic practices will grant big yields that feed the belly and the spirit. As part of the Connecticut-Northeast Organic Farming Association, they grow heirloom crops and raise pork and poultry humanely in pastures and woodlots. All meat is processed on-site and USDA inspected. Order pork and poultry from their website and vegetables from Healthy PlanEat (healthyplaneat.com). Muddy Roots partners with chefs and caterers to help them create tasty and beautiful dishes and reduce the transportation footprint. Visit them at the Stamford Museum & Nature Center's farmers market, Sunday, June through November.

River Crest Farm

534 Oronoque Rd., Milford; 203-876-9786; rivercrestfarm.com

Established in 1942 by the Acri Family, River Crest Farm is a member of the Northeast Organic Farming Association. The farm has grown and changed since Ralph Perry first cleared the Milford land for animals and vegetables, selling eggs and milk from his truck while wife, Irene, established the greenhouse. Now in the fifth generation, Maria, her husband, Andy, and their children keep their core values of "small, diverse, sustainable." They practice earth-friendly growing methods, including certified organic compost and organically approved fertilizers to grow vegetables, herbs, native and pollinator pathway plants, and even Christmas trees. Sign up for a CSA share for weekly pick up of fresh greens in June; beets, beans, and onions through July; eggplant and peppers in August and September; then pumpkins and popcorn as fall winds down. CSA members can also pick their own herbs and gather cut flowers. The greenhouse is open weekends 1 to 4 p.m. but check the website and blog for seasonal updates. Like their Facebook page for updates on farm cat sightings (#caturday).

Sherwood Farm

355 Sport Hill Rd., Easton; 203-268-6705; sherwoodfarm.org

Driving down Route 59 in Easton, you can't miss the giant rooster (or his little sidekick) who watches over Sherwood Farm. Bright yellow with red comb, this handsome guy isn't the only creature you'll find on the farm. Free-range chickens (of course) are joined by horses, goats, and cows. You'll also spot the modestly sized (for a dinosaur)

T-Rex in the pumpkin patch. Then look out for farm dog Lolly. The Sherwood family has been working this land since Noah Sherwood established the farm "about 1760," and it continues today as one of the oldest family farms in the US. CSAs are available, and the farm stand is fully stocked with Sherwood's freshly picked produce including greens and beans, eggplant and herbs, peppers and tomatoes, pumpkins and mums. Fill your basket with farm-fresh eggs, milk and yogurt, bread, bacon and local sausages. The stand is open Monday through Friday 10 a.m. to 7 p.m., and Saturday and Sunday 9 a.m. to 6 p.m.

Silverman's Farm

451 Sport Hill Rd., Easton; 203-261-3306; silvermansfarm.com

From July to October, Silverman's Farm is abuzz with pick-your-own patrons, all grabbing peaches, apples, and plums to eat fresh or to enjoy over the long winter. However, the consistent draw here is the 5-acre animal farm, which attracts countless children to its petting zoo every year. Silverman's has animals you'd expect to find on a farm: sheep, pigs, and goats (which walk across a fascinating raised boardwalk). But what delights the kids most are the buffalo, the emus, and the llamas. Silverman's also has a year-round country

store with homemade pies, cider donuts, jams, produce, salsas, and flowers, while the greenhouse features annuals, perennials, and holiday gift baskets. Visit the farm for a picnic, and let the kids play in the sand-like corn kernels and hay bales of the "Cereal Bowl." Take a tractor ride to the pumpkin patch at the annual Fall Fest, or sign up for camp or student tours. Closed New Year's Day to early spring; otherwise, the farm stand and animal farm are open daily 9 a.m. to 5 p.m.

Sport Hill Farm

596 Sport Hill Rd., Easton; 203-268-3137; sporthillfarm.com

Sport Hill boasts 40 acres of natural and nutrient-dense crops, with 100 varieties of non-GMO, sustainable produce, as well as chickens and pigs. That's an astonishing diversity—and that is the future of farming. Owners Patti and Al Popp believe in bringing seasonal, sustainable, and authentic food to the community and educating each

COURTESY OF SPORT HILL FARM

of us about why it matters. "We run all kinds of events on the farm for fun," says Popp. "But also to teach the value of good food and about what quality really is." That quality can be found in every bite of Sport Hill's colorful, vibrant, and healthy produce. Their seasonal market is open Tuesday, Thursday, and Friday 10 a.m. to 6 p.m., Saturday 10 a.m. to 5 p.m., and Sunday 10 a.m. to 4 p.m.

Stone Gardens Farm

83 Saw Mill City Rd., Shelton; 203-929-2003; stonegardensfarm.com

They say when you love what you do you never work a day. Fred and Stacia Monahan love what they do, and their hard work has made Stone Gardens Farm one of Fairfield County's largest vegetable growers. With 50 acres of fields, they practice Integrated Pest Management for healthy soil and healthy produce. They also raise pigs and cattle that are free from antibiotics, raised sustainably on locally sourced feed, and treated humanely. Chickens and turkeys are farm raised, too, and processed fresh. The farm stand carries their freshly harvested vegetables and meats, along with a variety of farm-made salsas, vinegar, dried beans and fruit, and baked goods. You'll also find dairy products and specialty items from local vendors and farms. Sign up for a chicken CSA for 10 weeks of chicken; beef shares (quarter, half, and whole) are also available. Join the Farm Credit program, similar to a CSA, which allows participants to get any items from the store year-round. The farm also features live music during the Summer Music Series. Celebrate the fall season with Halloween dinner. The farm store is open year-round Tuesday, Wednesday, and Saturday from 10 a.m. to 5 p.m.; Thursday and Friday from 10 a.m. to 6 p.m.; and Sunday from 10 a.m. to 4 p.m.

Stuart Family Farm

191 Northrup St., Bridgewater; 860-355-0172; stuartfamilyfarm.com

Food raised in a quality-controlled environment yields a healthier meal. The Stuart family understands this and practices this philosophy because it matters—for our satisfaction, for the good of the land, and importantly, for the lives of the animals who are in their care. The family has been doing it right for a long time (since 1929), letting their Red Angus cattle graze on 800 acres in Bridgewater, earning grass-fed certification from A Greener World (AGW), and Animal Welfare Approved (AWA) certification for their husbandry and environmental management practices. In addition to grass-fed beef, Deb and Bill raise free-range poultry and pastured pork, all available for affordable prices at the farm store and for home delivery. Bulk beef and pork orders are available for the next barbecue or for your backyard smoker. The farm store is open year-round Saturday 10 a.m. to 4 p.m. and Sunday 12 to 4 p.m.

Farm Breweries and Cideries

Stewards of the Land Brewery

418 Forest Rd., Northford; stewardsofthelandbrewery.com

With the convenience of packaging and artistic labels, it's sometimes easy to forget how much of what we love to eat and drink comes from the land. Or that transformations—like bread made from scratch or beer poured from the tap—are agricultural products. Farm breweries like Stewards of the Land want to change that. The DeFrancesco family, whose farm stand is right up the road in

Northford, have been stewards for five generations. President of the Connecticut Hop Growers Association Alex DeFrancseso cultivates 15 varieties of hops, uses water from the Farm River aquifer, and regionally sources malt to brew small batches of beer. Far exceeding the legal requirements for a farm brewery, their beer contains 80–90 percent locally sourced ingredients. Come by for a pour of an NEIPA like Yellow Fields, or Agave Rubio, a blond ale, and chat while overlooking the farm and grabbing a bite from Fire in the Kitchen or Mikro's Burger food trucks. Check the website for the food truck schedule, the lineup of music on Acoustic Sundays, and other special events. The taproom is open Thursday and Friday 3 to 11 p.m., Saturday 12 to 11 p.m. and Sunday 12 to 8 p.m.

Farmers Markets

Bethel Farmers Market

67 Stony Hill Rd; Bethel, CT; bethelfarmersmarket.org

The Bethel Farmers Market has been welcoming shoppers since 1981, one of the earliest ones in the state to do so. They operate on the grounds of the Fairfield County Agricultural Center on land leased by Stony Hill Preserves. Their vendors offer residents and visitors a great selection of local food. Veggies, fruit, dairy, and flowers come from Maple Bank Farm, Vaszauska's Farm, and Daffodil Hill Growers. Meat, sausages, and eggs are available from Eaglewood Farms and freshly harvested clams from Pepe's Cream of the Crop. Shop for specialty breads and sandwiches from Paradise Foods and Wave Hill Bread, or meat pies and "pasties" from Daily Fare. Other vendors include Goatboy Soaps and Marlene's Ariston Products, which carries infused oils, vinegar, and olives. Handcrafted wood sculptures, birdhouses, and other beautiful pieces are available from Wood You Believe It. The Bethel Farmers Market is active year-round: once a month January through April and every Saturday from 9 a.m. to 1 p.m. May through December.

Bridgeport Farmers Market Collaborative

Multiple venues, Bridgeport; bridgeportfarmersmarkets.org

Founded in 2014, the Bridgeport Farmers Market Collaborative is a community organization that supports multiple markets around the city that help residents get access to fresh, healthy food regardless of income. BFMC also allows independent organizations to pool resources for joint programming, fundraising, and marketing efforts. Member markets accept Supplemental Nutrition Assistance Program (SNAP) and WIC (Special Supplemental Nutrition Program for Women, Infants, and Children) benefits, as well as Senior Farmers Market Nutrition Program (SFMNP) checks, Bridgeport Bucks, cash, credit, and debit cards. Member markets operate late June or mid-July through October. East End NRZ Market and Café, located at 1851 Stratford Avenue, holds its market Sunday 9 a.m. to 3 p.m. St. Vincent's Hospital Farm Stand at 2800 Main Street meets Tuesday 11:30 a.m. to 4 p.m. in the medical center's parking lot. Wednesday from 10 a.m. to 2 p.m. is the East Side Market, located at the City's Social Services, 752 East Main Street. Thursday's markets are downtown at McLevy Green, the corner of Main Street and Bank Street, 10:30 a.m. to 2 p.m. and in the afternoon 3:30 to 5:30 p.m. at Bridgeport Hospital, 200 Mill Hill Avenue. On Saturday, the Green Village Initiative Reservoir Community Farm Stand, located at 1469 Reservoir Avenue, is open 10 a.m. to 2 p.m. Across the border, the Stratfield Saturday Farmers Market at Clinton Park runs 9:30 a.m. to 1:30 p.m.

CitySeed Farmers Market

Multiple locations, New Haven; 203-773-3736; cityseed.org

With so many great restaurants and food hotspots in New Haven, it's easy to forget that the bounty is out of reach for many, restricted by poverty, lack of transportation, and affordable choices for healthy food. Organizations like CitySeed help bridge the gap between food production and food access. CitySeed sows the kernels of community development, sponsoring markets throughout the city of New Haven. Collectively, the markets at Wooster Square, Edgerton Park, Fair Haven, and Downtown support economic growth for over 60 local vendors and provide access to healthy, affordable food year-round for more than 55,000 visitors. All sites accept Farmer Market Nutrition Program vouchers and SNAP and EBT cards.

Vendors at the various spots include Smyths Trinity Farm, Waldingfield Farm, Sugar Maple Farm, Four Mile River Farm, Riverbank Farm, as well as Olmo Bakery, Sono Baking Company, and The Traveling Farmer. You'll find salad greens, bagels, mushrooms, soup, and baked goods, all produced locally. Also sponsored by CitySeed, the Sanctuary Kitchen helps immigrants and refugees connect with their new home and build economic security and community involvement. You'll find their booth at the Wooster Square Market, located at 511 Chapel Street, on the site of Conte West Middle School. The Wooster Square Market is open April through December, 9 a.m. to 11 a.m. on Saturday. The market at Edgewood Park, at the corner of Whalley and West Rock Avenues, is open Sunday, June through November, 10 a.m. to 2 p.m. The Downtown Farmers Market takes place July through October on the New Haven Green across from City Hall at the corner of Church and Elm Streets, Wednesday, 11 a.m. to 2 p.m. The Fair Haven Market meets at the Quinnipiac River Park, Front Streets and Grand Avenue, July through October, Thursday 3 to 11 p.m. Check out the Winter Market in January and March and various pop-up venues for the Mobile Market.

Danbury Farmers Market

120 State St. (Danbury Railway Museum), Danbury; 203-792-1711; danburyfarmersmarket.org

The Danbury Farmers Market Community Collaborative (DFMCC) is a group of citizens and community sponsors who aim to bring healthy, nutritious, locally grown food to the community. They sponsor the seasonal farmers market and support the Better Food for Better Health campaign to enhance nutrition, sustainable agriculture, economic development, and to make sure fresh food is available for everyone. SNAP and WIC recipients and seniors can use their benefits at the market and the DFMCC will match Farmers Market Nutrition Program Vouchers used toward the purchase of nutritious food. Vendor offerings range from crafts and specialty items; baked goods and pastries; homemade jams, jellies, and salsa; along with farm fresh eggs, veggies, fruit, flowers, and meat. Vendors include farms like Killam and Bassette Farmstead, Clatter Valley Farm, Smith's Acres, Mitchell Farm, and Sullivan's Farm New Milford Youth Agency, which fosters agricultural knowledge and experience for young people. The market is open from the end of June to the beginning of November on Friday 10 a.m. to 2 p.m. with free parking at the Metro North Lot at the train station.

Downtown Milford Farmers Market

108 Main St. (Wasson Field), Downtown Milford; downtownmilfordfarmersmarket.com

The Downtown Milford Farmers Market began as an idea in 2005 put forth by members of the Downtown Milford Business Association. From that first seed, a collaboration with area businesses and farms has grown into an important community event that helps encourage economic and agricultural development for the benefit of all. Browse and buy fresh food from area farms, including Clover Nook Farm, Ekonk Hill Turkey Farm, River Crest Farm, and Vaiuso Farms. There are also prepared items and specialty foods from vendors like Oronoque Farm Bakery and Wild Woman Coffee; beauty, health, and wellness vendors; jewelry, home goods, and gifts. Live music and food trucks are part of the fun, too. The market meets Saturday 9 a.m. to 12:30 p.m., mid-June through the first week of October. A Senior Stroll begins at 8:30 a.m. for those 65 and over, pregnant, or immune compromised. Vendors accept Senior Farmers Market Nutrition Program (SFMNP) and WIC vouchers.

Farmers Market of Black Rock

St. Ann Church Field at 481 Brewster St., Bridgeport; farmersmarketofblackrock.com

The Black Rock section of Bridgeport is diverse and welcoming, with a hip art scene, funky eateries, tidal estuaries, and harbor views through Saint Mary's-by-the-Sea. The neighborhood's character is shaped by its history, landscape, and people. And every Saturday in summer and fall you can get fresh, locally grown and prepared food and locally made products at the farmers market. The nonprofit community program is run by volunteers and organized by the Black Rock Community Council. Seasonal fruits and veggies are available from vendors like Hungry Reaper Farm, Beckett Farm, and Oronoque Farms. Grab a loaf of sourdough, a crunchy-chewy baguette, or a few pretzel rolls from Fairfield Bread Company. Yoga in the Field begins at 10 a.m. ($10) and weekly live music will round out your day. The market operates Saturday 9 a.m. to 1 p.m., June to October.

Monroe Farmers Market

7 Fan Hill Rd. (Monroe Town Green corner of Rte. 110); monroefarmersmarket.org

The Monroe Farmers Market brings the community together to support farms, crafts people, and families for the common goal of supporting economic growth and providing access to locally produced food and specialty goods. With over 20 vendors, you'll find a wide range of items to select along with activities and events to keep you coming back. Farms selling fresh fruit and vegetables include Gazy Brothers, Laurel Glen, Waterview, and Hidden Gems Orchard. Upper Grass Greens offers specialty salad mixes and Oronoque Farms offers specialty baked goods. Breezy Knolls brings their all-natural beef products. Handmade goat-milk soaps and beeswax products come from Guardians Farm. There's also honey and maple syrup, all-natural granola, homemade guacamole, kettle corn, and Italian ice. You can buy wine from Sunset Meadows Vineyard and craft beer–infused goodies like Whisky Pretzels from The Drunken Alpaca. Weekly activities include entertainment for kids and their parents. Stop by the Hometown Table where nonprofit and other organizations set up and provide information and education about various community issues. The Monroe Farmers Market is open on Friday from mid-June to the third week of October from 3 to 6 p.m.

New Canaan Farmers Market

244 Elm St., New Canaan; newcanaanfarmersmarket.net

Located near the train station in the Lumberyard Lot on Elm Street, the New Canaan Farmers Market brings locals a great range of CT Grown fruits and vegetables, along with beautiful fresh flowers and herbs, tasty breads and baked goods, gourmet pasta, soup, cheese, shellfish, meat, and mushrooms. And so much more! Carrot Top Kitchens serve up fresh soups and salads, and you can get a healthful, probiotic brew from East Coast Kombucha or a superfood soft drink from Charlie's Changa Soda. Try Bees Knees Ice Pops, Michele's Pies, and Ivy Gourmet's granola. Ox Hollow Farm sells farm-raised beef, poultry, and eggs. Sustainable seafood is available from Copps Island Oysters and Ideal Fish. Calf and Clover Creamery offers raw milk and yogurt. Other farms bring the best of ripe, in-season as well as organic vegetables and fruit. Open May through the third week in November every Saturday 10 a.m. to 2 p.m., rain or shine.

Old Greenwich Farmers Market

38 West End Ave., Greenwich; oldgreenwichfarmersmarket.com

A visit to any market is a feast for the eyes and tastebuds. At the Old Greenwich Farmers Market you can sample so many things that local farms and business offer: squash blossoms, sunflowers, celery root, sweet corn, purple peppers, and purple cauliflower—seasonal, ripe, and healthy. The goodness comes from High Ridge Hydroponics, J&L Orchids, Riverbank Farm, Smith's Acres Farm, and Woodland Farms. Specialty vendors like Essentially Namaste from Cos Cob share natural health and wellness products. Beastly Threads offers eco-ethical textiles. Get a delicious Mediterranean dish from Yalla Organic Foods or a cup of restorative bone broth (with a side of dumplings) from Nit Noi Provisions. There's even something for your four-legged family from Yup Pup Treats. It's too much to pass up every Wednesday at the Living Hope Community Church, 2:30 to 6 p.m., May to November.

Shelton Farmers Market

100 Canal St., Shelton; sheltonctfarmersmarket.com

At the Shelton Farmers Market, you can stroll across from Riverwalk and Veterans Memorial Park and chat with vegetable and poultry farmers, then sample baked goods and breads made locally by area businesses. You'll find fresh produce from East Village Farm and Laurel Glen Farm; prepared foods from Lilu's Custom Catering and Liquid Lunch; and Three Bridges Coffee House; as well as healing elderberry soap from the Healing Herb Garden. The kids will love the Valley Sprouts Club, where they can learn about nutrition and eating, how to grow seeds, and why native pollinators are so important. Enjoy live music and food trucks, too, every Saturday from May through October, 9 a.m. to 12 p.m.

Trumbull Farmers Market

240 Unity Rd., Trumbull; niatrumbull.org/farmer-s-market

The Nichols Improvement Association hosts the Trumbull Farmers Market. This nonprofit organization sponsors a number of programs and events that help to preserve and promote the Nichols neighborhood of Trumbull. At the market you can find produce, fish, meat, and dairy from farms like Shaggy Coos, Eaglewood Farm, Why Not Farm, Too the Gills Fish Market, and others. Baked goods arrive from Saint-X foods, Sunnyside Sweet Shop, Pam's Cookies, and Biscotti Road. There's plenty of specialty from food vendors like Wonderland Jam, Hannah's Honey, and Dave's Angry Sauce. And if you're still looking for that unique "extra," grab a bottle of healthy kombucha or one of Continuum Distilling specialty spirits. Maybe a craft beer from Firefly Hollow Brewing Company or New Park Brewery. The Trumbull Farmers Market takes place May through October, Thursday 4 to 7 p.m.

Westport Farmers Market

50 Imperial Ave., Westport; westportfarmersmarket.com

Best known outside Connecticut for his acting genius, Westport resident Paul Newman also left a legacy of philanthropy. From Newman's Own salad dressings (proceeds going to charity) to the Hole-in-the-Wall Gang Camp for children with cancer, Newman gave back in so many ways. So, it won't surprise many to learn that the Westport Farmers Market sprang from his generous efforts, or that it started in the parking lot of the Westport Country Playhouse. In 2006, Newman (along with chef Michel Nischan) had a vision to support sustainable farming by bringing local producers together with residents. Since the market moved to a bigger space, along the Saugatuck River on Imperial Avenue, over 50 vendors join thousands of visitors each week. Growers like Fort Hill Farm, Popps Farm, Oxhollow Farm, and Woodland Farm offer fresh fruit, vegetables, honey, maple syrup, meat, and dairy. You can get shiitake and maitake mushrooms from Seacoast Mushrooms, seaweed from Stonington Kelp Farm, and oysters from Copps Island. And that's not counting prepared foods and specialty items like pizza from Dough Girls or gluten-free baked goods from Cloudy Lane Bakery. Other vendors rotate bringing you everything from nut milk and Nutty Bunny vegan ice cream to empanadas, vegan jerky, and craft brews. This vibrant market operates mid-May through early November, Thursday 10 a.m. to 2 p.m.

Festivals and Fairs

Guilford Fair

Guilford Fairgrounds, Guilford; 203-453-3543; guilfordfair.org

When the Guilford Fair began in 1859, farming was the staple of the community. Founding members of the Farmers and Mechanics Association, chatting at the local store on the west side of the green, conceived of a fair dedicated to farmers, working for improvements in agriculture and the "household arts." Today, the Guildford Agricultural Society sponsors this Library of Congress Local Legacy and honors that history, promotes local farms and farmers, and brings

COURTESY OF THE GUILFORD AGRICULTURAL SOCIETY

entertainment to the community. Every year at summer's end, the fairgrounds are full of festival goers checking out best-in-show pumpkins and curated vegetable baskets; saying hello to sheep, goats, and donkeys; snacking on fair fare; and waiting for the next performance of the Guilford Family Circus and live music. General admission ($10) covers the shows and exhibits with discounts for seniors and veterans. Three-day passes let you take advantage of everything, and children under 11 enter free, leaving more spending money for the rides and refreshments. Free parking and shuttle buses can be found at exit 57 off Interstate 95; parking on-site is $10. The fair takes place annually on the third full weekend of September, Friday 1 to 11 p.m., Saturday 9 a.m. to 11 p.m., and Sunday 9 a.m. to 7 p.m.

Milford Oyster Festival

Downtown Milford; 203-878-5363; milfordoysterfestival.org

Milford has been running the Oyster Festival since 1975, and it seems to get bigger every year. Today this annual August event is one of the largest in New England, with over 50,000 people attending. Lasting only one day (if you don't count the pre-event festivities), the festival features dozens of food vendors and hundreds of artists and craftspeople. Yes, hundreds. There are boat races, nationally known headlining bands, amusement park rides, cruises on the sound, and a classic car and motorcycle show. This is also a great time to check out the historic Wharf Lane complex with its three 18th-century houses (milfordhistoricalsociety.org). It's unclear whether the people come for all this great entertainment or the chance to consume dozens of local Milford-harvested, fresh, raw, on-the-half-shell oysters. Whatever the case, this is one event not to be missed if you're on the central Connecticut shoreline on the third Saturday in August. If you're not, change your schedule.

Norwalk Oyster Festival

Veteran's Memorial State Park, 42 Seaview Ave., Norwalk; 203-838-9444; seaport.org

Since 1978 people have been coming to Norwalk the weekend after Labor Day to celebrate (and eat) the mighty oyster. Among the clams, soft-shell crabs, and lobsters, you can prove your oyster-eating prowess in a "slurp-off" contest, or just try your first one. Run by the Norwalk Seaport Association, this festival is more than just food, with a focus on maritime heritage that includes old oyster boats, a shucking contest, marching bands, harbor cruises, arts and crafts exhibitions, and more. Check out the New England Village for arts and crafts and get your fill of fresh oysters, lobster rolls, or a traditional lobster dinner. Music is a big draw with local and big-name headliners to dance to all afternoon until the sun begins to set. Admission is $12 for adults, $8 for seniors, and $5 for children 5–12 years old. The three-day festival begins Friday 6 to 11 p.m., Saturday 11 a.m. to 11 p.m., and Sunday 11 a.m. to 7 p.m.

Orange Country Fair

525 Orange Center Rd., Orange; orangectfair.com

At the Orange Country Fair, there is fun for the whole family with pancake breakfasts and tractor pulls, doodlebug races, homing pigeon releases, and oxen and horse pull events. Browse all the exhibits and contests including baking (adult and junior categories), arts and crafts, livestock (cattle, goats, sheep, horses, swine, and llamas), birds of prey, snakes and reptiles, fruit, vegetables, and flowers. Get your hands dirty and test your strength (and guts) with the two-person handsaw contest, men's hay bale toss, and women's skillet toss. The fair operates on the third weekend in September, Saturday 8 a.m. to 7 p.m. and Sunday 8 a.m. to 6 p.m. Free parking and shuttle buses are available from remote locations. Admission is $8 for adults, $5 for seniors, and kids under 14 are free. Get a season pass through the Association of Connecticut Fairs (ctagfairs .org) and enjoy fairs across the state. What's a doodlebug? Come to the fair to find out.

Maple Sugarhouses and Apiaries

Monkey's Pocket Apiary

2788 Black Rock Turnpike, Fairfield; 203-371-4657; monkeyspocketapiary.com

Lining up against a stone wall, the custom-built hives of Monkey's Pocket Apiary are small colorful houses, a small neighborhood of hardworking bees. That's how the family sees it anyway. With the apiary in their back yard Kathy, Chris, and Morgan invite you to come pick up the honey from their kitchen table. Their products include wildflower, clover, buckwheat, and basswood honey, raw and kosher certified—full of immunotherapy properties, natural remedies, and local goodness. The family also shares their expertise—offering consultation, swarm retrieval and removal services, hive sales and set-up, and colony management help. Honey and other products can be found at Black Rock Pharmacy, In Touch Therapeutic Body Works, and other Fairfield County specialty shops. Sign up for the Adopt a Hive program and get honey from your very own bee hive. It's a great way to support these pollinators and get a personal supply of honey. Full and half-hive options are available. The apiary is open daily 9 a.m. to 5 p.m. Call ahead for pickup and look for the "Local Honey" sign and the yellow wheels of the farm wagon.

Warrup's Farm

11 John Read Rd., Redding; 203-938-9403; warrupsfarm.com

Warrup's Farm has been in operation for over 35 years. They're small in size but big in farming responsibility, selling farm fresh, local products—especially maple syrup, fall vegetables, and pumpkins. Visitors can see how syrup and cider are made, get their Christmas tree, visit farm animals, and take home something special. During the first three weekends in March, you can watch the wood-fired evaporator boil away, see the buckets on the trees, and visit the barnyard animals. Maple syrup and maple candy are also available for sale. Fall is prime time, with the pumpkin patch, hayrides, maze, and the farm store full of the harvest. Open seasonally, weekends 10 a.m. to 5 p.m.

Museums and Education

Dudley Farm Museum

2351 Durham Rd., Guilford; 203-457-0770; dudleyfarm.com

The Dudley Farm Museum and the Dudley Farm together make an important historical attraction and a great stop on your agricultural tour. The Dudley Foundation has restored this 1840s farmhouse (all 17 rooms) and runs the farm as it was in the 19th century. In the Munger Barn you'll find a display of Native American artifacts. And there's a sugarhouse, an herb garden, beehives, and animal enclosures with oxen, geese, and sheep. That's probably what the kids want to see. There's also an easy, kid-friendly 2-mile loop trail through the woodlands behind the farm. Tour the house and watch or participate in traditional farm and house chores. Experience maple sugaring on weekends during February and March, but call ahead, as these events are weather dependent. The admission-free museum is open from June through October, Thursday and Friday 10 a.m. to 2 p.m., Saturday 9 a.m. to 2 p.m., and Sunday 1 to 4 p.m. The best time to come may be on Saturday mornings when the Dudley Farm Farmers Market is held, beginning in late February through April, on the first and third Saturdays, 10 a.m. to 1 p.m. By May, the market is active every Saturday, 9 a.m. to 12:30 p.m. Local vendors sell fruits, vegetables, flowers, herbs, baked goods, jams and jellies, maple syrup, eggs, pickles, naturally raised meats, and handmade arts and crafts.

Ferreira Homestead and Apiary

22 Devonwood Drive, Waterbury; 203-293-7353; ferreiraapiary.com

Ferreira Homestead and Apiary has been raising bees for more than 100 years. The family has dedicated their efforts to helping others do the same. As educators they recognize, support, and promote the importance of bees and their essential role in nature. Without bees, humans don't eat, and bees need our help to combat the impact of climate change, habitat destruction, and disease. Master beekeeper Peter Ferreira shares his expertise through presentations, seminars, and mentorship programs. At the apiary, they maintain hives for demonstration and beekeeping instruction. Ferreira's additional bee yards and research apiaries contribute to the diversity of hives across the state. Their long-term goal is to increase the genetic diversity of bees and develop a strain that is resistant to disease. Ferreira provides hive maintenance consultations, mentoring and hands-on training, hive and swarm removal. Classes instruct hobbyists and professionals about bee anatomy and life-cycles, bee hierarchies and roles, and everything you need to set up and maintain your hives. Private and group classes are available for adults and children. The homestead is not open to the public, so email (beekeeper@ferreiraapiary.com) or call for information or to sign up for classes.

New Pond Farm Education Center

101 Marchant Rd., West Redding; 203-938-2117; newpondfarm.org

Set on 102 acres in the rolling hills of West Redding, New Pond Farm has been sharing their commitment to life-long learning and active engagement with the natural world for over 30 years. It is the vision of founder, actress Carmen Matthews, who trained at London's Royal Academy of Dramatic Arts, and then had a career spanning 55 years in theater, television, and film. With a passion for the environment and for education, she founded New Pond Farm in 1975 as a summer program for inner-city kids. The Learning Center building keeps the spirit of the original barn built in the 1700s, retaining antique, hand-hewn beams of America's early forests. Participate in an array of enrichment activities in the outdoor classrooms of woodlands, wetlands, and meadows. Join the cows, sheep, chickens, and honeybees at this working farm. Programs include astronomy lessons, farm demonstrations and hands-on practice, art shows, barn dances, and authentic Native American encampments. Summer programs bring children together and give them a chance to learn about farming through a shared experience. New Pond has also been recognized as a Connecticut Dairy Farm of Distinction; you can pick up delicious whole milk and chocolate milk from the Annex. The farm is open daily, 9 a.m. to 4 p.m., and the Annex is open daily 7 a.m. to 7 p.m. Register ahead of time for programs, and become a member to enjoy special events.

Stamford Museum & Nature Center Sugar House and Cidery

151 Scofieldtown Rd., Stamford; 203-977-6521; stamfordmuseum.org

The Stamford Museum & Nature Center is a destination that offers cultural and educational exhibits and activities. Their mission is to inspire life-long learning through interactive experience, and to build community knowledge, promote environmental stewardship, and nurture curiosity and creativity. The grounds include Knobloch Farmhouse, Heckschler Farm, Bendel Mansion, Overbrook Nature Center (with trails and Nature's Playground), and Wheels in the Woods (universally accessible trail). In other words, you'll be busy, and will have to plan multiple return trips for the observatory and planetarium, or to see the otters. At Heckscher Farm, you can be farmer for the day, feeding and grooming animals. The Sugar House showcases the science and art of maple syrup production throughout February and March, in preparation for the annual, three-day Maple Fest. At the Farm Festival Weekend in spring, enjoy sheep shearing, hayrides, and cute baby animals. Sunday farmers markets run June through November at the Knobloch Family Farmhouse parking lot (at the north entrance of 151 Scofieldtown Road). There you'll find farm-fresh produce, meat and dairy from local vendors, specialty food and baked goods, artisan products, as well as food trucks, storytelling, and a play zone for the kids. The museum and nature center are open year-round, except holidays, Monday through Saturday 9 a.m. to 5 p.m. and Sunday 11 a.m. to 5 p.m. Heckscher Farm is open daily April to October 9 a.m. to 5 p.m., and November through March 9 a.m. to 4 p.m.

Wakeman Town Farm

134 Cross Highway, Westport; 203-557-6914; wakemantownfarm.org

"A community farm with heart and soul," Wakeman Town Farm Sustainability Center is dedicated to serving the community, promoting education and sustainable farming and lifestyle practices. One of the oldest properties in Westport, this land was owned by eleven generations of the Wakeman family. At the turn of the 20th century, John Wakeman established a dairy farm with 40 cows and home milk delivery. Over three generations, the family expanded operations, eventually selling their 41 acres to the town in 1970. They retain a lifetime lease and have remained active in the land's stewardship, opposing some proposed developments and compromising elsewhere. Wakeman Park, dedicated in 1996, is one such compromise that established ball fields without overdeveloping former farmland. In 2009, after the passing of Irene Wakeman, the Westport Green Village Initiative (GVI) began leasing the farm and farmhouse, then renovated, and established the demonstration homestead. Today the Sustainability Center provides a number of activities and educational opportunities, while the Aitkenhead family lives on the property and takes on the role of primary ambassadors of sustainable, healthy living, continuing the tradition of stewardship. After all, this is a working farm, and everyone gets a chance to contribute. Visitors feed goats and pet chickens, Volunteers get a chance to get hands-on experience working the land, tending fields and animals, and living sustainably. In the gardens they grow leafy greens, carrots, beets, herbs, fruit, and more, all available at the farm stand. Young people gain life-long agriculture appreciation and potential career planning through after-school and summer camp programs and internship opportunities. Kids of all ages enjoy CSA pick up, farm dinners, Family Fun Day, Eco Market, Lectures on Sustainability, and so many great programs. Visit the farm and farm stand on Saturday 9 a.m. to 1 p.m.

Pick Your Own

Beardsley's Cider Mill and Orchard

278 Leavenworth Rd., Shelton; 203-926-1098; beardsleyscidermill.com

Beardsley's began as a dairy farm in 1849, but an unfortunate school bus accident and fire destroyed the barn in 1973. So, the family directed their attention to other crops and worked to keep the land safe from development. Today, their orchards grow apples, peaches, nectarines, and plums. Their dwarf apple trees are pruned to maximize air flow and allow the sun to reach every apple. There are over 16 apple varieties, all available for pick-your-own from September through October. After picking, you can make your own hard cider, choosing your perfect blend; Beardsley's will press it for you, and fill your fermenting containers. The farm store carries freshly baked pies, cookies, cider donuts (of course), maple syrup, gifts, and local raw honey from their own hives and others around the state. You can also pick up your cider-making kit. Open mid-September through Christmas Eve, 10 a.m. to 5:30 p.m.

Blue Hills Orchard

141 Blue Hills Rd.; Wallingford; 203-269-3189; bluehillsorchard.com

The Henry family has been tending to the fields and trees of Wallingford for six generations. William Henry left his post as the Dean of the College of Agriculture at the University of Wisconsin and journeyed to Connecticut, unable to resist the land that would become Blue Hill Orchard. They first farmed cabbage, peaches, and cherries, moving to apples by 1941. Today, they span 300 acres, harvesting over 250 acres of apples and additional acres of peaches, plums, and nectarines. The family also partners with a Massachusetts craft cidery to produce crisp and balanced Stormalong Hard Cider. Private groups and school trips can be arranged for farm tours and hayrides. The pick-your-own season starts in August, Saturday 10 a.m. to 5 p.m. and Sunday 11 a.m. to 3 p.m. The farm store is open Friday and Saturday 10 a.m. to 6 p.m. and Sunday 11 a.m. to 4 p.m., August to November.

Blue Jay Orchards

125 Plumtrees Rd., Bethel; 203-748-0119; bluejayorchardsct.com

In the 18th and 19th centuries, Connecticut was absolutely blanketed in apple orchards, from which people made a delightful, mildly alcoholic cider drunk by adults and children alike. Today there are only a few farms left in the state (and especially in Fairfield County) with large apple orchards, and Blue Jay in Bethel is one of them. In fact, it was the first farmland preserved by selling development rights to the state, guaranteeing it must always stay a working farm. This is good news for us, since it's true what mom used to say: an apple a day keeps the doctor away. From August to October, you can enjoy this treasure through pick-your-own apples, of which they have dozens of varieties. The pumpkins come right on schedule, from September through Halloween, so visit the patch after you've filled your apple basket. The Bakery Store is open 11 a.m. to 5 p.m. until December, with gifts and baked goods, including, of course, apple pie.

Drazen Orchards

251 Wallingford Rd., Cheshire; 203-272-7985; drazenorchards.com

David Drazen founded these orchards in 1951 when he purchased land, moved his family, and dug in, tending to trees and harvesting McIntosh, Macoun, and other varieties. The farm has remained a family affair, now under the care of farmer-in-chief Gordon, and children Eli and Lisa and their kids and dog. The orchard implements Integrated Pest Management and compact trellising practices to manage peaches, pears, plums, quince, and apples—including Cortland, Jonagold, Honeycrisp, Zester, and others. Take a wagon ride to pick your own or stop by the farm store for Drazen's harvested produce (fresh or in a pie) as well as products from other local farms. Blueberries are ripe in July and August, followed by peaches, nectarines, pears, and plums. Apples start in August with Ginger Gold and finish with Golden Delicious in October. Call ahead for updates on what's ripe in the trees. The farm stand is open late July to November, daily 9 a.m. to 6 p.m.

Hickory Hills Orchards

363 South Meriden Rd., Cheshire; 203-272-3824; hickoryhillorchards.com

Hickory Hills Orchards has been the passion of the Kudish family for over 40 years. First started in 1977, Fred and Lynn purchased the land without intending to farm it themselves. Soon enough, they discovered that doing it themselves meant they could do it right. They passed their business sense and stewardship philosophy to their two daughters and son-in-law, who grow apples, peaches, pears, plums, and nectarines on 33 acres. The Hickory Hills store carries harvested fruit, baked goods, locally produced artisanal gifts, honey, preserves, and decorations. Of course, after apple picking, you'll want to pick up cider donuts or apple fritters. Pick-your-own season starts in mid-August, and the orchards are open every day through November 9:30 a.m. to 5 p.m. The store is open until 6 p.m.

Norton Brothers Fruit Farm

466 Academy Rd., Cheshire; 203-272-8418; nortonbrothersfruitfarm.com

Come to Norton Brothers Fruit Farm for a picnic, to pick your own, or pick up some fresh produce at the farm store. The Norton Family has been farming in Cheshire since the mid-eighteenth century.

Seven generations of brothers, fathers, sons, mothers, daughters, and aunts have kept the farm operating, making sure blueberries, apples, pears, and peaches are ripe and delicious year after year. With blueberries ripening in July, nineteen varieties of apples, and six types of pears, you'll have your pick of something delicious all summer long. At the farm market, you'll find Norton's harvest along with locally sourced seasonal vegetables and fruit, like native strawberries, salad greens, and corn on the cob. In addition to fresh produce, enjoy jams, jellies, maple syrup, honey from orchard bees, fresh and frozen pies, and homemade cider and donuts, muffins, and fritters. Order gift baskets and seasonal goodies for Thanksgiving and Christmas. The store is open June through December, Monday through Friday 8 a.m. to 6 p.m. and Saturday and Sunday 9 a.m. to 5 p.m. Pick-your-own hours are 8:30 a.m. to 5 p.m. Monday through Friday and weekends 9 a.m. to 4:30 p.m.

Rose Orchards

33 Branford Rd., North Branford; 203-488-7996; roseorchardsfarm.com

Start your visit with a bite of the past: Robert Rose arrived from England in the early 1600s, and the family has been part of Connecticut's rich history ever since. Today the eleventh and twelfth generation of Roses still live on and care for 50 acres of crops and animals. Visit the barn and say hello to goats, ducks, and chickens. On the farm they grow a variety of vegetables from corn to cucumbers, as well as fruit, most of which you can pick yourself. Pick your own strawberries, raspberries, peaches, pears, and apples 9 a.m. to 5 p.m. in season. In addition to PYO, Rose Orchards features a market, grill, and creamery. The market offers bakery-fresh pies, in fourteen scrumptious types, from apple raspberry to strawberry rhubarb to pumpkin. Rose's Creamery frozen custard starts with chocolate or vanilla then creates something magical—twists, Razzles, sundaes, and shakes. Add fresh seasonal fruit to make it even better. The market is open seven days a week, 9 a.m. to 6 p.m, year-round. Buck's Grille serves breakfast and lunch every day 9 a.m. to 3 p.m. The creamery is open seasonally 9 a.m. to 6 p.m.

Sea Farms and Aquaculture

Copps Island Oysters

7 Edgewater Place, Norwalk; 203-866-7546; coppsislandoysters.com

The folks at Copps Island Oysters operate on tradition. Norm Bloom and Son have been oystermen since 1940s, and still do it with old-fashioned know-how and passion, ensuring sustainable crops and delicious harvest of oysters from the cold, nutrient-rich waters of Long Island Sound, across the Connecticut coast. With a fleet of 15 ships and a crew of hard-working guys and gals, and a philosophy of stewardship, they are committed to farming healthy oysters and maintaining healthy waters. Wild-grown Copps oysters from Norwalk are sweet and briny. From the mouth of the Mystic River, come Mystics, balanced with mineral notes. Ram Island oysters taste, as oysters should, like the ocean. Norm Bloom and Son offers wholesale and direct-to-chef shellfish, but at the shop you can purchase your own, along with little neck clams and lobsters, shucking equipment, sportswear, beach gear, and specialty products. The Norwalk store hours are Monday through Friday 9 a.m. to 5 p.m. and Saturday 9 a.m. to 4 p.m.

Indian River Shellfish

178 Cottage Rd., Madison; 203-605-5158; indianrivershellfish.com

Oystering has been a staple of life in the waters of Long Island Sound off the Connecticut coast since early colonists learned the practices of coastal tribes. But over-fishing and pollution took its toll over the centuries. The good news is that today oyster farming is a growing industry that contributes to the health of the water and the sustainability of the mollusks. The aquafarmers at Indian River Shellfish Farm started small, learning on the job, cultivating their passion into a viable industry—the hard way with hard work and trial and error. George Harris and Mike Gilman sold their first harvest in 2012. Since then, they've expanded to an 11-acre farm, raising hatchery-spawned Eastern Oysters in the waters off Clinton and Madison, and "dabbling" in naturally set wild oysters. Order online

and pick up at the dock on Cottage Road Friday 12 to 4 p.m., or Saturday 9 a.m. to 12 p.m. Get shucking!

Thimble Island Ocean Farm

CSF only; thimbleislandoceanfarm.com

Thimble Island Ocean Farm started the first Community Supported Fishery (CSF) in the country, selling shellfish and seaweed. In this way, consumers participate in regenerative farming that keeps the ocean healthy. Buying CSF shares works the same way as agriculture shares work. You get fresh seafood at a reasonable price, and collectively the financial contributions allow Thimble Island to seed shellfish beds, maintain hatcheries, and grow and harvest seaweed. Shellfish shares offer monthly deliveries (June through November) of oysters and clams. Seaweed shares include whole and baby leaf sugar kelp or specialty products like pickled kelp, kelp bagels (from Olmo restaurant in New Haven), kelp burgers, or kelp dolmades (from Tavern on State in New Haven). Since the COVID-19 pandemic, shares also support GreenWave, a nonprofit co-founded by Thimble Island Ocean Farm founder and award-winning chef Bren Smith. The organization works to educate ocean farmers, support other small commercial aquafarms, and promote a healthy aqua-ecosystem. Sign up online and pick up CFS shares from New Haven on the first Saturday of the month.

Trail Rides

Blue Spruce Farm

626 Wheeler Rd., Monroe; 203-268-6774; bluesprucehorseriding.com

Blue Spruce Farm offers family fun for horse lovers and riders of all levels. Lessons focus on holistic practices drawn from tai chi and yoga that help riders connect with their animal and learn proper techniques. Guided trail rides take you through scenic woodlands of Monroe, navigating through summer's green and autumn's changing colors, with an optional wade in the nearby pond. Connect with horses in a supportive setting and learn from experts who help novices and seasoned riders feel safe and comfortable. Blue Spruce is a great option for summer picnics, birthday parties, and summer camp. Open daily 10 a.m. to 5 p.m.

Gold Rush Farms

5 Silver Hill Rd., Easton; 203-268-9994; goldrushfarmsct.com

Family oriented and devoted to horses and people, Gold Rush Farms offers a unique experience for all riders. The farm is situated on 130 acres in Easton, offering lessons, boarding, leasing, and training, as well as trail rides and summer camp activities. Riders 8 and up can enjoy a guided expedition though the farm's private, wooded trails. Rides last for 45 minutes ($50) or 90 minutes ($90). Pony rides around the farm and pony parties are perfect for the little ones. Trails operate year-round, Wednesday through Monday from 9 a.m. to 5 p.m. Call early to schedule your ride; they book up quickly.

Rap-A-Pony Farm

995 North Farms Rd., Wallingford; 203-265-3596; facebook.com/rapapony

Catering to beginner riders, Rap-A-Pony Farm has been teaching children (of all ages) English and Western riding techniques. They have school vacation and summer programs for very reasonable prices if you want to learn the care and riding all in four days. Otherwise, they have one-hour group lessons, where they teach you how to tack and ride properly, or half-hour private lessons (which of course can be extended). Rap-A-Pony has indoor and outdoor arenas, so you can even go on a rainy day or in cooler weather. They provide helmets but be sure to wear sturdy shoes and clothes. Open Saturday 9 a.m. to 3 p.m. Summer camp, party, and group sessions are also available.

Rowanwood Farm

175 Huntingtown Rd., Newtown; 203-270-8346; rowanwoodfarm.com

Trail ride adventures aren't just for horse lovers, and you don't need to put your butt in a saddle to have fun outdoors, connect with animals, and support CT agritourism. What a better way to explore than to hike with a mini llama at Rowanwood Farm! Meet the family of Peruvian, Bolivian, Chilean, and Argentinian mini llamas and let them escort you through the green and russet foliage of rural Newtown. Hikes include a brief introductory Llamas 101 session and a chance to connect with your new buddy, then a guided trek. Hikes are suitable for beginners ages 8 and up, and last one and a half hours with plenty of stops for picture ($50 per person). For more experienced hikers and return visitors, you can navigate with your guide (human and four-legged) for 2 hours, through moderate to challenging terrain ($80 per person). Cria Crawl Farm Tours offer the chance to explore the farm's Lakota tipi and interact with other rescue animals. Suitable for all ages, families, and groups, the tour is one hour ($30 per person, children under 2 are free). Other events at the farm (30 Chestnut Hill Road, Sandy Hook) include healing drum circles the first Friday of every month, which allow you to tap into the energy and rhythm of Mother Earth.

Tree Farms

Broken Arrow Nursery

13 Broken Arrow Rd., Hamden; 203-288-1026; brokenarrownursery.com

Broken Arrow began in 1947 when Dick Jaynes replaced apple orchards with Christmas trees, growing to the thriving 20 acres that you can visit today. The nursery was added in 1984, making it a full-time business where Dick used his 25 years of experience at the Connecticut Agriculture Experiment Station to cultivate mountain laurel and other shrubs, plants, and trees. Over 1,500 varieties are now propagated, grown, and sold at the nursery, and they've expanded acreage and opened a second location in Hamden. Broken Arrow grows 24 varieties of broadleaf evergreens, 10 types of deciduous shrubs, evergreen conifers, and herbaceous perennials. Cut-your-own or pre-cut Christmas trees include Balsam, Concolor, and Fraser Firs, along with White, Norway, and Colorado Blue Spruce. Handmade wreaths with traditional greens, winterberry, and holly come in a variety of sizes. The nursery is open for sales daily April 5 through October 31, from 10 a.m. to 5 p.m., by appointment in November and March, closed January and February. During Christmas season (November 20 through December 24), they open the satellite location at 680 Evergreen Avenue.

Castle Hill Farm

One Sugar Ln., Newtown; 203-426-5487; castlehillfarm.net

Established in 1927 as a dairy farm, Castle Hill Farm now offers plenty of seasonal agritourism activities. On their 100 acres, the family-run farm grows Christmas trees, hay, corn, pumpkins, and other vegetables. In summer, visit Meme's Market farm stand for fresh vegetables, eggs, honey, and grass-fed beef and pork from down the road at Kearns Farm. There are also farm-to-table dinners and opportunities for special events at the Cedar Pavilion (planning a wedding?). The fall is a great time to visit the corn maze and stroll the sunflowers, pick your own pumpkins, and go for a hayride. Part of Castle Hill is Paproski's Tree Farm (at 5 Hattertown Road). As the holidays approach, you can stroll the 42 acres of Douglas Fir,

Blue and White Spruce and cut your own. Pre-cut trees as well as wreaths, holiday arrangements, and other goodies are also available. The farm stand is open July through October, Saturday and Sunday 10 a.m. to 5 p.m. The Sunflower Stroll opens in September and the 7-acre corn maze, pumpkin patch, and hayrides are open in October and November. The tree farm operates weekends from Thanksgiving to Christmas, 9 a.m. to 5 p.m.

Fairview Tree Farm

2 Sawmill City Rd., Shelton; 203-944-9090; fairviewtreefarm.com

Christmas, tradition, and family go together, and Fairview Tree Farm champions all three. Shelton native Bradley Wells established the tree farm and landscape construction business in 1989 on land farmed by his family since the 1840s. With 60 acres to explore, you can meet the seedling helpers (dogs Moose and Gryffin), visit the goats, gather pumpkins, and enjoy walking trails and hayrides. In October don't miss the Legends of Fear Halloween attractions including haunted hayrides and spooky trails (if you dare) through the farm's woods. Check the event calendar: tours start late September through Halloween. After Thanksgiving, 'tis the season to take a ride with Santa then browse the Christmas Shoppe filled with decorations, ornaments, and trimmings as well as with hot cocoa and cider donuts. Open during the holiday season 9 a.m. to 5:30 p.m. every day except Tuesday.

Maple Row Farm

555 North Park Ave., Easton; 203-261-9577; mrfarm.com

Maple Row Farm has been farming in Easton since 1769. Originally a dairy farm, they began growing Christmas trees in the 1920s, inspired by a 4-H project of young Sherwood Edwards. Since then, the Edwards family has kept the traditions of sustainable farming and land stewardship with Christmas trees, managing 10-year growing cycles, firewood, and mulch production. They grow 9 varieties of spruce and fir trees on 200 acres, but don't worry, you can get a tractor ride back to your car. When you visit, be sure to check out the teams of oxen (Bill and Joe; Mike and Pete) and think back to how farms operated with the help of these animals who could pull

COURTESY OF SABIA TREE FARM

up to 5,000 pounds! In addition to Christmas trees, the team offers a range of landscaping services. The farm is open March through October, Monday through Saturday 7:30 a.m. to 4 p.m. The holiday tree season starts late November until Christmas, open daily 9 a.m. to 4:30 p.m.

Sabia Tree Farm

772 Morehouse Rd., Easton, 203-650-4429; sabiatreefarm.com

In late autumn, when our attention turns once again to the holiday season, this is about the time to get to Sabias Tree Farm, roam the well-tended rows of spruce, and find the one that's just right. You can borrow a handsaw and their crew will help you get it on the car. Pre-cut trees and pin stands are also for sale. There's even a fire going to warm up. More than trees, though, you'll find a great selection of handmade wreaths, glass ornaments, maple syrup, rustic signs, Connecticut honey from Monkey's Pocket Apiary, and teas from Whole Harmony in Haddam. Outside the Christmas season, you can select produce from a 100-year-old wagon transformed into the summer farm stand by Backwood Builders in Newtown. The stand is open April to November and works on the honor system. The tree farm is open from Thanksgiving to Christmas, starting on Black Friday (10 a.m. to 5 p.m.), then Saturday and Sunday 9 a.m. to 5 p.m.

Van Wilgen's Garden Center

51 Valley Rd., North Branford; 203-488-2110; vanwilgens.com

Nursery salesman William C. Van Wilgen immigrated from Holland in 1920 and began propagating nursery stock on the property near Old Pine Orchard Road, including a variety of rhododendron cultivated by his brother called "Wilgen's Ruby," which is still available today. Since then, four generations of the Van Wilgen family have helped grow the business with wholesale nursery stock and a landscaping company. Today, they continue to do it all, operating greenhouses and garden marts in Milford, Guilford, and Old Saybrook (and coming soon to Clinton), as well as the main store in North Branford. They specialize in many varieties of trees, shrubs, tropicals, citrus, perennials, roses, fruit, and houseplants. Their experts will help you select the right tree for your yard, grown to optimize root development and make it easy to transplant to your soil or planter. Whether you're looking for herbs and vegetables for the backyard or a perfect Turkish Fir for under the Christmas star, you can find them at the garden center in North Branford, open daily 8 a.m. to 5 p.m.

Vaszauskas Farm

519 Middlebury Rd., Middlebury; 203-758-2765; vaszauskasfarm.com

Like so many other farms in Connecticut, Vaszauskas doesn't fall neatly into one category. Trees are just one element of this 300-acre farm in Middlebury. Founded in 1950, the family-owned plant nursery grows a variety of plants, vegetables, and evergreens. At the garden center you'll find trees, wreaths, garlands, and roping available for all your Christmas and winter decorating plans. Evergreens include White and Cedar Pine, Concolor, Normand, Balsam, Fraser, and Blue Spruce. The garden center also features annuals, perennials, shrubs, hanging baskets, mulch, and other landscaping products. At the farm stand, open every day 8 a.m. to 7 p.m., you'll also find fruits and vegetables, rounding out the farm's offering. Don't miss the corn—butter 'n sugar and silver queen are local favorites.

Wineries

Aquila's Nest Vineyards

56 Pole Bridge Rd., Newtown, 203-518-4352; aquilasnestvineyards.com

The founders and owners of Aquila's Nest Vineyard, Neviana and Ardian, draw upon their Albanian heritage and myths from their culture and others, like ancient Greece and aboriginal Australia. Neviana's last name, Aquila, is Latin for "eagle," and Homer tells of Aquila, who carried Zeus's thunderbolts and found his place for him in the stars as Aquarius. Ardian was born under the sign of Aquarius and remembers his grandfather telling stories from his time in Australia about how the Corona Borealis is seen as an eagle's nest, a place to gather under and share stories. At the vineyard, they bring these traditions to winemaking and hope to share these deep roots with customers. They're raising their sons on the 40-acre farm where they have planted 10 acres of eight grape varietals, including cool-climate hybrid reds like Chambourcin, Marquette, and Baco Noir, and whites Cayuga, Traminette, and Valvin Muscat. They're also growing *vinifera* Cabernet Franc and Riesling. Their intriguing lineup of wines is a class in world mythology; enjoy Queen of Illyria (a red blend), Princess of Troy (Merlot), the Amazons (Baco Noir), Sybil (a dry rosé), Siren (Moscato), and Zane e Malit (a dry Riesling named for an Albanian fairy). Hours: Thursday 6 to 9 p.m., Friday 5:30 to 9 p.m., Saturday 11 a.m. to 9 p.m., and Sunday 11 a.m. to 8 p.m. Revel in the sunset views from the tasting room and share your own tales with friends and family.

DiGrazia Vineyards

31 Tower Rd., Brookfield; 203-775-1616; digraziavineyards.com

Founded in 1978, this winery offers a unique array of wines. Founder and winemaker Dr. Paul DiGrazia pioneered the production of wines rich in antioxidants (try the Wild Blue—made with blueberries) and has created some of the most interesting blends on the Connecticut Wine Trail. In the tasting room you find whites featuring grapes like Traminette, Seyval Blanc, and Vidal Blanc. Red blends, blush-style, and dessert wines are also featured, including a spectacularly smooth port-style wine and another that blends pear, black walnut, and brandy. Local honey can be found along with wine, pumpkin

cinnamon, ginger, nutmeg, and cloves in Autumn Spice or the slightly sweeter Harvest Spice. Open weekends 11 a.m. to 5 p.m.

Gouveia Vineyards

1339 Whirlwind Hill Rd., Wallingford; 203-265-5526; gouveiavineyards.com

Established in 1999, this winery brings the Portuguese heritage of its owner, Joseph Gouveia, to the Wallingford Hills. The tasting room is a gorgeous stone structure, and visitors flock to the site year-round. Inside, high wood beams frame the larger of two rooms, greenhouse-like windows shape the side walls, and a double-sided fireplace warms patrons in winter. In warmer months, you're likely to see picnickers on the lawn and on the patio under the pergola, sipping Stone House Red and sharing cheese, sandwiches, or possibly a birthday cake. The winery is equally beautiful in winter, as late afternoon light reflects off the snowy vines. Gouveia offers a wide range of wines, including whites like Seyval Blanc, Cayuga, and two styles of Chardonnay. For a red, try the Cabernet Franc, which is rich and earthy. Open daily 11 a.m. to 8 p.m. and Sunday 11 a.m. to 6 p.m.

Jones Family Farms and Winery

605 Israel Hill Rd., Shelton; 203-929-8425; jonesfamilyfarms.com

The Jones family has farmed their land among the White Hills of Shelton for over 150 years. More than 400 acres, the farm is a local favorite in any season, where visitors can discover first-hand the family's commitment to the land. Christmas trees were planted on the dairy farm in the 1940s by fourth-generation son Philip. A few decades later, his son Terry added strawberries and blueberries, while Terry's son Jamie created the winery in 2004. Bring the family and pick your own berries in June and July, or enjoy the autumn leaves or a light snowfall as you pick out your perfect pumpkin or select the best tree. But save time for a visit to the tasting room, with its beautiful, handcrafted bar, Thursday to Sunday 12 to 6 p.m. Whites like Pinot Gris and Stonewall Chardonnay are crisp, so they pair well with foods. Red offerings include Cabernet Franc—one of the best examples in the state—and Jones's signature blend, Ripton Red, that has notes of vanilla and spice. The Harvest Kitchen cooking studio did not operate during the pandemic, but you can trust that it will be back. Check their website for current opening hours as well as for all activities and events. At Jones you can see how "Be good to the land, and the land will be good to you" is a credo we can all adopt. They've shown how through seven generations.

Paradise Hills Vineyard

15 Wind Swept Hill Rd., Wallingford; 203-284-0123; paradisehillsvineyard.com

You'll feel as though you stepped into an Italian landscape painting when you arrive at Paradise Hills. The owners have been growing grapes for over 25 years and opened the Tuscan-style tasting room in 2011. The ambience, inside and out, is elegant and airy, and you're sure to admire the copper bar and patio as you relax and sip the small but varied selection of reds and whites. The historic Washington Trail passes through the property, and a red and a white blend are both named for it. Varietal-named favorites like Merlot, Cabernet Sauvignon and Marquette can be found on the menu, as well as Italian blends like Nero D'Avola and Sapor Dolce. The old world meets the new Monday through Saturday 11 a.m. to 8 p.m. and Sunday 11 a.m. to 6 p.m.

Rosabianca Vineyards

1536 Middletown Ave., Northford; 203-208-1211; rosabiancavineyards.com

When Andrea Rosabianca immigrated to America, he brought with him his love of family and the sharing of great wine. While raising children, working as a carpenter, and tending his gardens, he passed on these traditions, making wine and working hard. In

2010, Andrea's son Charlie fulfilled his father's dream of opening a vineyard. Today, the family owned and operated winery sits on the slopes of Northford in a restored red-barn tasting room, where you can enjoy a flight of Italian-style whites and reds under the majestic maple on the veranda. Their Estate grown offerings include Dolce Chicco D'oro, made with Seyval Blanc grapes, as well as Dolce Gocce, which showcases Cayuga White grapes in a semi-sweet floral white. In addition to two versions of Bianca Rosato rosé—in both dry and sweet styles—you'll want to sample (and probably take home) one of their reds or red-blends, like Sangiovese and Tramonto. Raise a glass of Vino Del Nonno, "My Grandfather's Wine," a full-bodied red, smooth and bold with notes of cherry. Open year-round except holidays. Check out their website for tasting room hours, and food truck and farm dinner events.

Savino Vineyards

128 Ford Rd., Woodbridge; 203-389-2050; facebook.com/savinovineyards

Framed by stone walls and located in a quiet neighborhood in Woodbridge, Savino Vineyards might go unnoticed, except for the 8 acres of vines that sprawl beside the driveway. The family-owned-and-operated winery was opened in 2009 by Jerry Savino, a native of Salerno, Italy, and his wife. The small red tasting room is quaint, with a bar, vintage cash register, and a few tables, where you can sample the Seyval Blanc, aromatic with citrus and mineral notes, or the dark plum Frontenac, with notes of cherry and vanilla. Share a plate of antipasto as you sample St. Croix; check out live music Sunday in the summer or events like Paint & Sip or Yoga & Wine. Open Saturday and Sunday 12 to 6 p.m., mid-May to mid-December.

Umpawaug Farm Winery

299 Umpawaug Rd., Redding; 203-493-1428; umpawaugwines.com

Being small can be a good thing. After all, Umpawaug Farm Winery attributes their success in producing award-winning estate wine to a hands-on approach to winemaking and vineyard management. With 5 acres on three locations in Redding, they grow cold hardy red varietals Chambourcin, Marquette, Corot Noir, Marechal Foch,

Frontenac, and Cabernet Franc. In other words, "This is not your father's Cabernet." And that's the great part of winemaking in Connecticut. By choosing grapes that do well in Western Highlands, and by limiting production, they get results. The folks at Umpawaug also carefully manage the landscape; grape skins are composted, wood cuttings are mulched, and vine canopies are thinned to control sun and airflow. Grapes are actually harvested at night to maximize brix levels (how much sugar will be fermented into alcohol). The results in the glass are bold flavors and balanced tannins, with layered undertones, and spicy finishes. Try Corot Noir if you're looking for something more fruit-forward, or Marechal Foch for an earthy middle and smokey finish. The winery is currently open Saturday and Sunday 1 to 6 p.m. for bottle sales, but check for future hours.

White Silo Farm and Winery

32 Rte. 37 East, Sherman; 860-355-0271; whitesilowinery.com

White Silo Farm and Winery specializes in wine made from fruit grown on the property. The winery and tasting room are found in an old barn; and, yes, it features a white silo. The rhubarb wine will surprise you, and you'll fall in love with the raspberry, blackberry, and black currant wines. If you thought fruit wine was always sugary and powerfully sweet, try these made in the dry style. If sweeter is your style, and even if it's not, White Silo offers dessert versions as well. Try the black currant Cassis or savor the puckering goodness of Raspberry and compare it to the dry version. Pick raspberries and blackberries on-site and enjoy the picturesque family-run farm. The tasting room is open weekends, April to December: Friday 11 a.m. to 6 p.m. with extended hours to 8 p.m. during the summer; Saturday and Sunday 11 a.m. to 6 p.m.

COURTESY OF ARETHUSA FARM DA

NORTHWEST CONNECTICUT

The charming peaks and dales of the Litchfield Hills and the wet valleys of the Connecticut River are a center of agritourism, with many educational opportunities and showcase farms. Driving through the tiny village of Salisbury or the sprawling metropolitan suburbs around the state capitol, you happen upon farm after farm, many started by 19th- and 20th-century immigrants who made this area home. The shade leaf tobacco farms of the rich, flat floodplains north of Hartford have long been the suppliers of the wrappings for Cuban cigars, and the Western Connecticut Highlands AVA is completely within this region. Indeed, some of the greatest Cabernet Francs in America can be tasted here.

The Litchfield Hills have also pioneered the natural food movements, with a grocer like New Morning Market in Woodbury operating for over 50 years and bringing organic food to customers. That pioneering spirit is necessary to maintain a family farm, diversifying and modifying for future generations of consumers. "Connecticut's independent food producers persevere," says Dana Jackson, publisher and editor of *Edible Nutmeg*, who on days off from the magazine can often be found doing farm work in the northwest hills. "Fusing sensible agricultural traditions with requisite remodeling for modern challenges."

Arethusa Farm

556 South Plains Rd., Litchfield; 860-567-8270; arethusafarm.com

Arethusa Farm was saved from development in 1999 and has since become one of the most successful dairy farms in Connecticut. Once the old barns were renovated and 300 cows brought in, the forward-thinking owners built their own small dairy plant in the Bantam firehouse. Not content with that, they opened a restaurant, Arethusa al tavolo, in 2014, along with a bakery and several retail stores creating a food empire that stretches across the Nutmeg State. Once you taste their products, you'll know why they are so successful. They allow fresh cream to sour for two days before churning small-batch cultured butter. They make nine aged cheeses and 16 percent butterfat ice cream. Oh, and their gardens and greenhouses produce year-round fruit and vegetables, including apples, pears, plums, apricots, and several types of berries. Heirloom salads, tricolor beets, and sungold cherry tomatoes are grown for their restaurant tables, or yours, when available seasonally in their stores. The Bantam retail store is open Sunday to Thursday 10 a.m. to 8 p.m. and Friday to Saturday 10 a.m. to 10 p.m. The Arethusa a Mano cafe is open every day but Tuesday 7 a.m. to 4 p.m., while their restaurant is open Sunday, Wednesday, and Thursday 4:30 to 7:30 p.m. and Friday and Saturday 4:30 to 8 p.m.

COURTESY OF ARETHUSA FARM DAIRY

Collins Creamery

9 Powder Hill Rd., Enfield; 860-749-8663; facebook.com/Collins-Creamery-262357045256

Standing outside the Collins Creamery, we heard someone remark, "Why is all the best ice cream made at farms?" Let's hope it was a rhetorical question. Farms are so clearly the best place to get ice cream, for five reasons at once, but we'll answer it in the way it was at the time: freshness. The Collins Creamery makes over 20 flavors at their small dairy store at Powder Hill Farm, and you can get them in waffle cones, sundaes, and milk shakes, along with a variety of other vehicles, every day of the week 11:30 a.m. to 9:30 p.m. But of course, what you want is the delicious, creamy substance that has made gluttons of us all. They also have yogurt, sorbet, and fat-free and soft-serve ice cream. In Connecticut, many dairy farms were not able to compete with giant corporate farms out west that are as big as our entire state. Many, like Collins, have turned to gourmet ice cream, and we can all be glad they did.

Griffin Farmstead

30 Copperhill Rd., East Granby; 860-716-3239; griffinfarmstead.com

One of only a few goat dairies in Connecticut, Griffin raises registered Nubian, Saanen, and alpine goats on pasture. They milk them seasonally April through November, and make a yogurt, cheese, and soap, along with bottling milk and chocolate milk. They also grow flowers and veggie starters in the spring and mums and pumpkins in the fall. Recently, they also purchased some shorthorns for milking, and you'll see alpaca wandering around the fields, as well. The Griffin family also owns Clark Farms in Suffield—see the entry below—and sells their goat milk there, as well as at many other stores in the area. Call for hours at the Griffin Farmstead.

Hastings Farm

472 Hill St., Suffield; 860-668-1061; hastingsfamilyfarm.com

This multi-generational farm was once part of the great tobacco farm belt in the Connecticut River valley. Howard Hastings decided wisely to diversify and added a dairy herd. His children and

COURTESY OF TRENA LEHMAN

grandchildren and great-grandchildren expanded the dairy and recently expanded the range of dairy products and added natural beef. The 100 Holsteins and Jerseys on the farm are milked twice daily, and you can stop in and meet them and their adorable calves. Along with milk and chocolate milk, their Greek-style small-batch yogurt comes in vanilla, coffee, blueberry, lemon, and many other flavors. They craft Cheddar, Gouda, and Havarti, sold by the pound. Open Monday through Saturday 9 a.m. to 6 p.m., the store also carries other Connecticut farm products, from Maple Valley ice cream to Ekonk Hill turkeys.

Kimberly Farm

415 Chestnut Land Rd, New Milford; 860-354-1839; kimberlyfarm.com

The Kimberly family have run this dairy farm since 1955, producing delicious whole creamline milk in several flavors. They also make yogurt, drinkable yogurt, mozzarella, and eggnog. At their store, open daily 9 a.m. to 6 p.m., you can also buy farm fresh eggs, various cuts of beef including skirt steak and picnic roasts, and seasonal produce from lettuce to eggplants. Their exquisitely flavored ice cream is obviously a draw, with waffle cone swirl, candy cane, and toasted almond flavors drawing applause.

Lost Ruby Farm

458 Winchester Rd., Norfolk; 860-542-5806; lostrubyfarm.weebly.com

With milk from their free-ranging, pasture-raised Saanen goats, micro-dairy Lost Ruby crafts pasteurized goat cheeses from traditional European recipes. "Proud to be small, slow, and inefficient!" is the rallying cry of owners Antonio Guindon and Adair Mali. Their aged Black Pearl is mixed with garlic, salt, and smoked paprika, giving it a unique smoky flavor, while their Graystack uses fine wood ash to moderate the acidity, giving it a rich complexity unheard of in most American dairies. They also raise Buff Orpington and Black Sexlink chickens for egg production (the bright orange yolks are beautiful and delicious), as well as pigs and turkeys. Check their website or call ahead for farm pickups Thursday through Sunday at 2 p.m. and 6 p.m.

Smyth's Trinity Farm

4 Oliver Rd., Enfield; 860-745-0751; smythstrinityfarm.com

This family owned and operated dairy farm in Enfield is open Monday through Friday 6 a.m. to 6 p.m., and Saturday 6 a.m. to 4 p.m. They pasteurize and bottle their own milk in beautiful glass bottles right on the farm and you can take a tour to see it happen, as well as visiting their grass-fed cows. They also produce yogurt, butter, and cream, and sell honey and other delights at their store. You can find their milk at other farms around the state, and Smyth's even offers

home delivery to Enfield, Somers, Suffield, Windsor Locks, Longmeadow, and Ellington. It is part of a new . . . or is it old . . . movement to bring back milk delivery in the 21st century. It's a great way to support your local farm, and also to make sure you always have fresh milk.

Thorncrest Farm and Milk House Chocolates

280 Town Hill Rd., Goshen; 860-491-4261; milkhousechocolates.net

Have you ever thought about the flavor of chocolate made from an individual cow? At Thorncrest you get that opportunity. "Chocolate should be just as fresh as tomatoes," says Kimberly Thorn. The farm harvests their own hay and pastures their cows, ensuring the cows are fed the finest, sweetest hay and natural feeds. And it's a good thing they take such care, because from each cow, single-cow-origin chocolates are crafted. The shop is open Monday through Saturday 10 a.m. to 5 p.m. and Sunday 10 a.m. to 4 p.m. You can also visit the cows that produce this remarkable milk Thursday through Sunday 10 a.m. to 3 p.m.

Tulmeadow Farm Store and Ice Cream

255 Farms Village Rd., West Simsbury; 860-658-1430;
tulmeadowfarmstore.com

The Tuller family has been farming in West Simsbury since 1768, and like many farms in Connecticut, this one has diversified extensively. They have cattle, vegetables, and an ice-cream store, which they began in 1994. From mid-April through October, you can stop by from 9 a.m. to 9 p.m., with the ice-cream window opening at 12 p.m. You can shop at their farm store for their own produce, like lettuce, arugula, chard, tatsoi, basil, tomatoes, cucumbers, peppers, scallions, broccoli, cabbage, bok choy, sweet corn, squashes, zucchini, cucumbers, green beans, and eggplant, as well as their grass-fed beef. They also carry the products of other local farms, like milk, honey, maple syrup, bread, salsa, chips, pesto, goat's cheese, fudge, tomato sauce, bruschetta, and Nodine's Smokehouse meat. And, of course, you can try one of their 50 flavors of award-winning 16 percent butterfat ice cream, the most popular of which might be red raspberry with chocolate chips.

Farms and Farm Stands

Anderson Farms

165 Broad St., Wethersfield; 860-529-2662; andersonfarmsct.com

Founded in 1854 by James Anderson, this farm used to be part of the famous Wethersfield onion trade, as well as growing tobacco and raising dairy cows. Anderson made enough money through agriculture to build one of the gorgeous homes on the town's Broad Street Green. The farm remains in the family today, selling a diverse crop of traditional New England vegetables, like fresh spinach,

COURTESY OF TRENA LEHMAN

radishes, and sweet corn. The stand on the green is open seven days a week 9:30 a.m. to 5 p.m., May through October.

Averill Farm

250 Calhoun St., Washington; 860-868-2777; averillfarm.com

This farm dates back to colonial days, and the Averill family has been farming the land since 1746. The 200 acres boast structures that date to the early 1800s, including the huge old stone homestead. Ten generations of Averills have worked the land, and they continue to grow 100 varieties of apples and pears, which are available for picking. The farm stand is open Labor Day to Thanksgiving, and in addition to their own fruit, they sell cider, donuts, maple syrup, honey, cheese, and cut flowers. Friday to Sunday 9:30 a.m. to 5:30 p.m. you can get their fresh-baked pies, but be quick about it, because they sell out. The pear-cardamom crumb pie should be illegal it is so good, but traditionalists who want the pumpkin pie will also be delightfully surprised by the ginger crumb crusts and the maple roasted walnut topping.

Belltown Hill Orchards

483 Matson Hill Rd., South Glastonbury; 860-633-2789; belltownhillorchards.com

Donald and Mike Prelli are the third generation of the family to own the farm on Belltown Hill, begun by their immigrant grandparents in 1910. They have diversified so much, it is difficult to put Belltown into a category, despite the "orchards" part of their name. As you might guess, you can pick your own peaches, pears, cherries, apples, pumpkins, and tomatoes, or buy the finished products in one of their great pies at the bakery. You'll also find apple cider donuts, caramel apples, jams, jellies, and more. However, picking out your Christmas tree or taking educational field trips on the farm in a private wagon are options, as well. The Prelli farm is open every day 9 a.m. to 5 p.m. June through October, and usually Monday through Saturday in November and December. You can often catch them in January on weekends only. Belltown makes a secret-recipe applesauce that has become famous in the region; get some and try to figure out what they put in it to make it so darn good.

COURTESY OF TRENA LEHMAN

Broad Brook Beef

47 Broad Brook Rd., Broad Brook; 860-250-3311; broadbrookbeef.com

This Connecticut ranch raises natural, pasture-raised, dry-aged beef and pork. For four decades they have specialized in British breeds of cattle for enthusiastic beef connoisseurs and local "foodies" who are looking for a local place to buy premium, restaurant quality, hormone free meat products. Their herd is not large, limited on purpose to keep the level of quality high. In 1978 when the current owner's parents, Herb and Kathy Holden, purchased the farm, this idea was way ahead of its time. Today, we care a lot more about where our food is raised and how it was treated. Can you imagine even a decade ago caring so much that their Berkshire hogs have access to rooting pens? So, Broad Brook's time has come. Their farm store is open Saturday, 10 a.m. to 3 p.m. all year long, and their CSA farm share program runs from October to March, with steaks, chops, and everyday meats in six 20-pound boxes, along with all-natural soup mixes. "We're from New England and we know what New Englanders expect when it comes to their food," say the folks at Broad Brook. And that means quality.

Brown's Harvest

1911 Poquonock Ave., Windsor; 860-683-0266; brownsharvest.com

The Brown family have been farming in and around Windsor, Connecticut, since the mid 1800s. Six generations have grown a variety of crops including potatoes, strawberries, and Connecticut Valley Shade Tobacco. Since 1978, Brown's Harvest has participated in the community by providing education for families and groups from schools, clubs, and businesses. Their 175 acres are planted with sweet corn, asparagus, pumpkins, strawberries, and blueberries, available at their seasonal stand Wednesday through Friday 11 a.m. to 6 p.m., and Saturday and Sunday 10 a.m. to 5 p.m. You'll also find fresh baked goods, honey, and, yes, ice cream. On weekends in the autumn, they have hayrides from 10 a.m. to 4:45 p.m. and sponsor a number of festivals and workshops throughout the season. Have you ever tried a flashlight maze? The cider donuts at the end taste like sweet freedom.

Clark Farms at Bushy Hill Orchard

29 Bushy Hill Rd., Granby; 860-413-9733; clarkfarmsct.com

Many stop at Clark Farms for the popular cider donuts, or the ice-cream parlor, but this multi-use 78-acre farm includes a cidery, cafe, and orchard. The 15,000 pick-your-own apple trees are joined by peaches, blueberries, Asian pears, nectarines, plums, and a variety of vegetables. You'll even see beehives in the orchard. The full bakery does not only make their piping hot cider donuts, but a full range of made-from-scratch pies, apple dumplings, turnovers, cinnamon buns, tea breads, and scones. The sweet cider, hard cider, and apple wine produced by the cidery are all blended from fruit on the farm; you can grab a growler to go. The cafe and cidery are open Thursday and Friday 3 to 8 p.m., Saturday 9 a.m. to 8 p.m., and Sunday 9 a.m. to 5 p.m. They also have a greenhouse at 676 Mountain Road in Suffield, 860-668-4086. Check their event schedule for farm to table and farm to families breakfasts and dinners.

D'Agata's Fine Family Farm

1448 North Grand St., West Suffield; 860-668-6906 or 860-670-2311; dagatasfinefamilyfarm.weebly.com

At the far north border of the state, this first-generation family farm raises beef, pork, turkeys, and chicken and offers free-range eggs and native vegetables like potatoes and acorn squash during the season. They also offer butchering services for wild game. Stop by at their farm stand, open daily from 7 a.m. to 5 p.m. and pick up fresh roasting chickens or a side of bacon for the fridge.

Dondero Orchards

529 Woodland St., South Glastonbury; 860-659-0294; donderoorchards.com

Celebrating more than 100 years of history, the farm on Woodland Street was established in 1911 by immigrants from Italy, Joseph and Mary Ann Dondero. Their grandson Joe, his wife Sandy, and their children operate the farm today. It is known for its pick-your-own,

with veggies, blueberries, apples, and more. You'll also find the Dondero crew at many of the region's farmers markets, though it has two farm stores of its own, open every day 9 a.m. to 4 p.m. from March to December (until 5 p.m. in the summer). Held every other Wednesday at 6:30 p.m. from May to October, their farm dinners have different themes, from clam bake to steak night, from BBQ to Harvest Moon. They also boast a 7,200-square-foot greenhouse, offer a popular CSA, run a bakery, and make their own pickles, jams, and pesto. Always looking for opportunities to improve, Dondero's expanded with the acquisition of the Mountain View farm location (3582 Hebron Avenue). Oh, and one more thing. Their holiday pies. They make a lot of pies: apple, apple crumb, blueberry, blueberry crumb, fruit of the farm, fruit of the farm crumb, (apples, raspberries, blueberries, blackberries, and strawberries), peach, pecan, peanut butter apple crumb, pumpkin, strawberry rhubarb, chocolate cream, coconut cream, Key lime, and lemon meringue. But even though they make so many, and so many different kinds, they somehow run out. So, make your holiday order early and often, and put it on the calendar for the following year, too.

Eaton Farm

55 Red Oak Hill Rd., Farmington; 860-284-8239; eatonfarm55.com

Since 1955 this family farm has been serving the Farmington community, using biodynamic and organic growing practices. From late May to Thanksgiving their farm stand is open daily 7 a.m. to 7 p.m., and they sell firewood year-round. They recently started a CSA program, offering four different options and providing 22 weeks of produce to members. Check out their eight (!) varieties of garlic.

Fair Weather Acres

1146 Cromwell Ave., Rocky Hill; 860-529-6755; fairweathergrowers.com

For over a century, the Collins family has been farming the rich soil of the Connecticut River Valley. Once dairy farmers, they have become primarily vegetable farmers these days, with an astonishing variety (over 200) available in their large CSA program or at their farm store from May 1 to Christmas, open every day 9 a.m. to 6 p.m. Their lunch "kitchen" is open 11 a.m. to 2 p.m., serving delicious wraps and sandwiches. They have 16 acres dedicated to their Fall Festival and Corn Maze, which also features a hayride, scarecrow

COURTESY OF TRENA LEHMAN

walk, grain corn pit, tire playground, pedal go-karts, and much more. This is at their other address at 101 County Line Drive in Cromwell (actually just down the road, but across the town border). It is open in late September and October, Friday to Sunday 10 a.m. to 6 p.m., though the last ticket is sold at 4 p.m. to give you time to solve the maze!

Farm Faraway Homestead, LLC

267 Hard Hill Rd. North, Bethlehem; 203-232-5360; farmfarawayhomestead.com

This husband-and-wife team provide the community eggs, honey, starter plants, and vegetables, including squash, peppers, cucumbers, beans, peas, and a variety of greens. The small stand at the house is open for self-serve every day during the summer and fall, 7 a.m. to 7 p.m. You might even meet their chickens, who roam naturally on the farm, living their best life.

Freund's Farm Market & Bakery

324 Norfolk Rd. (Rte. 44), East Canaan; 860-824-0650; freundsfarmmarket.com

Theresa Freund and family began their farm stand as "a door propped up on 2 blocks holding a bushel basket of sweet corn."

COURTESY OF CABOT CREAMERY COOPERATIVE

Now it is one of the premier farm stores in the state, with seasonal produce like tomatoes, peas, beans, corn, and squash from their 400 acres or products from other regional farms, from Cabot cheeses to Nodine's Smoke House meats. They pickle, jam, and gumbo their produce, as well, and make an amazing variety of pies (try the egg custard!). The bakery alone is worth a stop, with pastries, biscuits, donuts, cookies, and dozens of breads from ciabatta

COURTESY OF CABOT CREAMERY COOPERATIVE

COURTESY OF CABOT CREAMERY COOPERATIVE

lunga to braided challah. You can also join the Freunds for a "stroll" on the land, walking and talking about agriculture and conservation, grazing and enjoying the fruits of their labor. They are open in the summer Monday through Saturday 9 a.m. to 6 p.m. and Sunday 9 a.m. to 5 p.m.

Gresczyk Farms

860 Litchfield Turnpike, New Hartford; 860-482-3925; gresczykfarms.com

"Gresczyk-grown" may not be easy to say five times fast, but it is a motto in the New Hartford area, nevertheless. Their crops include everything from corn to cucumbers, potatoes to pumpkins, and branch out into hydroponic lettuce and year-round herbs. They have 600 hens laying eggs in graded sizes, and they make their own marinaras, salsas, jellies, and fruit butters. The farm store also sells CT Grown fruit and meat, and Connecticut-made syrup, yogurt, and much, much more. Along with 130 acres of vegetables, the Gresczyks have nine greenhouses that grow plants from pansies to mums, available from early spring through the autumn. You can join their CSA if you're in the area, or stop by their farm store from Easter to Christmas, Monday through Saturday 9 a.m. to 7 p.m. and Sunday 9 a.m. to 6 p.m.

Hard Rain Farm

57 Spielman Hwy. (Rte. 4), Burlington; 860-675-3941; facebook .com/Hard-Rain-Farm-103292288119115

Tom Roberge of Hard Rain Farm is something of a local legend, appearing like a phantom at local farmers markets and surprising newbies with his unbelievably fresh, crisp vegetables. Author David K. Leff calls his "brilliant orange" carrots the "world's best," and Tom's eggs, apples, onions, and maple syrup are not far behind. Small farmers like Roberge keep agriculture hyper-local and laser-focused on quality. You can stop by the farm during the summer and early fall and buy his products on the honor system or find him and his food by chance, like all the best things in life are found at the northwest Connecticut markets.

Holcomb Farm

113 Simsbury Rd., West Granby; 860-844-8616; holcombfarm.org

We first encountered Holcomb Farm through its public trails; wedged between Granby Land Trust and the McLean Game Refuge, it is part of a network of farmland and woodland that comprises one of the premier walking areas in the state. They are open 365 days from dawn to dusk and are fantastic in the winter for snowshoeing and cross-country skiing, as well as horseback riding and hiking all year round. Their "tree trail" is an arboreal education in itself. However, this 312-acre historic farm is so much more than that. Farmed by the Holcomb family for more than two centuries, it was given to the University of Connecticut and then passed on to the Town of Granby, which owns it today. Run by a nonprofit called Friends of Holcomb Farm, it operates a CSA program and farm store, both of which offer a mixed selection of farm vegetables, from black radish to padron peppers, from carrots to kohlrabi. The store also features locally sourced artisanal foods from other farms, bakeries, and dairies from the region, and is open mid-June to late October, Tuesday through Saturday 10 a.m. to 6 p.m. This is community agriculture at its finest.

Jillybean's Farmstand

172 Scott Swamp Rd. (Rte. 6), Farmington; 860-839-5887; jillybeansfarmstand.com

Opening Easter weekend and closing on Christmas, this 100-acre family farm grows corn, tomatoes, eggplant, potatoes, and strawberries. Pick up their own produce or fresh brown chicken eggs at their farm stand, along with peaches, plums, apples, and blueberries from neighboring orchards. It is open daily from Easter weekend to Christmas Day, 9 a.m. to 5 p.m. Their CSA sign-ups start in February, and during the holiday season they have Christmas trees, wreaths, and gifts. Their buttery sweet corn may be the best in the state.

Karabin Farms

894 Andrews St., Southington; 860-620-0194; karabinfarms.com

"Know your farmer, know your food," says Diane Karabin. And getting to know Karabin Farms is an education on how a diverse farm operates, with greenhouses, Xmas trees, an apple orchard, a pumpkin field, a maple syrup shack, and so much more. They have flowers and plants galore and an amazing selection of beef cuts and turkey parts, as well as some of the most delightful duck and goose eggs we've had. Like many farm stores, they also have a selection of goodies from other Connecticut farms: ice cream and milk, maple syrup, and honey. You can pick your own apples, everything from Zestar to Ginger Gold to Snow Sweet in September and October, Saturday and Sunday 9:30 a.m. to 4 p.m. They have pick-your-own pumpkins during the same months, and same times. Then in November you can get pre-cut or pick-your-own Christmas trees, up to 8-foot firs and spruces, available every Saturday and Sunday until sold out. And in the winter? February and March mean maple syruping, with weekend demonstrations of the process, so call ahead. The farm store is open daily April 1 through December 23, 9 a.m. to 5 p.m., with January through March on Saturday and Sunday 9 a.m. to 5 p.m.

Killam & Bassette Farmstead

14 Tryon St., and 1098 Main St., South Glastonbury; 860-833-0095; kbfarmstead.com

This 1893 family farm specializes in all-natural chicken, eggs, and pork, available in many cuts and styles, from shoulder roasts to Polish kielbasa. Their canned goods (particularly their pickles) are top-notch, and they sell dozens of varieties of produce from blackberries to cauliflower, from garlic to grapes, from snap peas to summer squash. They feature "Fresh from the Farm Tastings" in the spring and fall, with free samples of farm-grown goodies, local coffee, live music, hayrides, a jam tasting bar, face painting, kids activities, a maker's market, live animals, and more. You can also join their CSA, pick your own flowers and pumpkins, and shop for other Connecticut-made goods at the store. Their 14 Tryon Street stand is open daily on the "honor system," but if you'd like their meats, call ahead and order it before you go so it will be ready for you. The other address, 1098 Main Street, South Glastonbury, also operates as an honor system stand. Both are open daily 9 a.m. to 6:30 p.m.

Ox Hollow Farm

478 Good Hill Rd. (Rte. 317), Woodbury; 860-354-3315; oxhollowfarmct.com

A diversified family farm nestled in Roxbury, Ox Hollow straddles three towns, with its barn and farm store on Good Hill Road and another farm stand at 1474 Litchfield Road (Rte. 202) in Bantam. Since 1994, Mark Maynard II has been raising hormone and antibiotic free, all-natural, pasture-raised Angus beef, pork, lamb, and poultry. Using movable coop enclosures and rotational grazing, he raises broiler hens and also offers seasonal field turkeys. The barn store at Good Hill Farm in Woodbury is open May to November 11 a.m. to 6 p.m.

Percy Thomson Meadows

78 Thomson Rd., Bethlehem; 203-266-5785 or 203-598-9701; percythomsonmeadows.com

This fifth-generation, 100-acre farm once focused on dairy, but today specializes in all-natural, grass-fed beef, pork, chicken, and lamb. You can join their Farm Membership Share to guarantee yourself a season's worth of meat, or visit their farm store, where in addition to prime cuts, they also have their own smoked and processed meats, prepared by Noack's of Meriden, including bacon, frankfurters, bratwurst, kielbasa, knockwurst, liverwurst, and ham. Their whole chickens are huge and juicy, and their eggs are all guaranteed fresh. For those of us who also want a few veggies, they have plenty, including Swiss chard, beets, Brussels sprouts, and many more. The farm stand is at 98 Main Street South in Bethlehem and operates 365 days a year on a self-service honor system, or you can make your order online through their website and pick it up.

Rosedale Farm

25 E. Weatogue St., Simsbury; 860-651-3926; rosedale1920.com

Just down the road from the mighty Pinchot Sycamore, Rosedale Farm is committed to providing Simsbury and the region with fresh fruits, vegetables, flowers, and even wine. Established in 1920, the farm is now in its fifth generation. With four varieties of sweet corn and at least that many kinds of tomatoes, the farm market is stocked

with beautiful, healthy produce—from beets to broccoli, carrots to cabbage, potatoes, peppers, and parsnips, just to name a few. Nine species of flowers are grown for summer bouquets, and you'll also find local cheeses, breads, jams, and cakes in the market. Diversification is one of the keys to the family-owned operation, and in 2005 they opened a winery. Award winners include Three Sisters, an Alsatian-style white made from estate-grown Cayuga white blended with Niagara grapes. Chef-to-farm dinners are a summer staple on the farm, and corn mazes and hayrides will delight young and old in autumn. The live music every Saturday and Sunday 1:30 to 4:30 p.m. during the season makes this farm what it should be—a community hub that brings the whole town together.

Southwind Farms

223 Morris Town Line Rd., Watertown; 860-274-9001; southwindfarms.com

In 1998, Jim and Penny Mullen founded this unique farm, at first raising sheep but then turning to alpacas. Today over 50 of these curious and gentle animals roam 23 acres, producing some of the highest quality alpaca wool in America. Call ahead to make an appointment for a tour of the farm and pick up some yarn while you are at it. Their store is open from late autumn to the day before

Christmas with clothing, teddy bears, and gift items, though their yarn is available throughout the year in 15 natural, undyed colors and 70 dyed colors. They hold events throughout the year, like the Daylily Alpaca Fest in July and the annual Fall Festival in October, so check their website for dates. They also sell alpacas to the right buyers, so if you are in the market for an investment rather than a sweater, this is the place to come.

Stuart Family Farm

191 Northrup St., Bridgewater; 860-210-0595; stuartfamilyfarm.com

Henry Stuart bought this property back in 1929, and today the Stuart Family Farm continues his traditions, with some modifications for modern consumers. "The majority of our clients express to us that their biggest concern when purchasing meat is if the animal was humanely raised and how well the animal was treated during slaughter," says Bill Stuart. As a certified grass-fed AGW farm and certified Animal Welfare Approved farm, the Stuarts are doing it right. Their animals are raised outdoors and on pasture for their whole lifetime, eating grass and foraging. Visit their farm stand to buy this remarkable meat Saturday 10 a.m. to 4 p.m. and Sunday 12 to 4 p.m. and other times by appointment.

Sub Edge Farm

199 Town Farm Rd., Farmington; subedgefarm.com

Located right between the Farmington Golf Club and Avons Old Farms School, Sub Edge Farm was named by famed architect Theodate Pope Riddle after she bought the 100-year-old dairy farm and used it to educate the students at the school. Today, Rodger and Isabelle Phillips grow vegetables, fruits, flowers, and culinary herbs, as well as raise heritage breed pigs,

pasture-raised chickens, and grass-fed beef. They have a CSA to join and work closely with restaurants in the Farmington area, providing them with specialty produce. Look at their schedule on the website for their popular farm dinners with the DORO Restaurant Group. The farm store on the idyllic road (check out the cows by the giant Robert Frost tree) is open Tuesday through Saturday, 10 a.m. to 6 p.m., and Sunday 10 a.m. to 3 p.m.

Sun One Organic Farm

50 Maddox Rd. Bethlehem; 203-266-7973; sunoneorganic.com

Sun One is not only an organic farm, free of pesticides, herbicides, or artificial fertilizers, they believe that hand tools and dedicated workers are the best way to "reconnect with nature and grow the healthiest food possible." The farmland has been in the Maddox family since the 1800s, and current owner Rob is committed to local, organic produce. They host events and farm tours to show you the amazing machine-free work they do and have a CSA to join to bring this healthy, nutrient-dense food to your family. Their farm store is open Saturday 9:30 a.m. to 1 p.m. during the season, and offers their own produce as well as coffee, cookies, beauty products, and more. They also feature a small 165-square-foot geodesic dome (built from wood from the property) listed on airbnb that you can rent out if you need a space to really get away from it all.

Truelove Farms

296 Thomaston Rd., Morris; 203-217-6234; truelovefarms.org

"A pig doesn't want to live in a shed," says Tom Truelove. "Because they are allowed to lead their lives entirely outdoors and in our woodlot pastures, they get a lot of exercise and are naturally healthier." Truelove's heritage-breed hogs are also free of growth hormones and sub-therapeutic antibiotics. At this 100-acre farm's store you can also get chicken, eggs, turkey, and beef from Friday to Monday between Memorial Day and Columbus Day. "If we eat three meals a day," says Truelove. "We have three opportunities to support our community, tangibly demonstrate our values, and rejoice in the wonder of good food!"

Waldingfield Farm

24 East St., Washington; 860-868-7270; waldingfieldfarm.com

This certified organic-growing farm owned by the Horan brothers has recently expanded from heirloom tomatoes to apples, maple syrup, and various vegetables. Patrick Horan likens going to your local farm as "like joining the local YMCA or joining the library," stressing that Waldingfield is a "community" that works only when we participate. They must be doing something right, because their CSA has over 200 member shares. You can also get their produce from a farm stand, open Memorial Day to October, 12 to 4 p.m. Tuesday to Sunday, ¼ mile past their main driveway.

COURTESY OF WALDINGFIELD FARM/DANA JACKSON

Whippoorwill Farm

189 Salmon Kill Rd., Lakeville; 860-435-0289; whippoorwillfarmct.com

The folks at Whippoorwill have been raising marbled, grass-fed beef for decades, aging it, and then flash-freezing it in cryovac to keep a consistent quality. They sell a lot to local restaurants, but you'll find individual cuts of this perfect beef at their farm store year-round, Friday and Saturday 10 a.m. to 5 p.m. (or call and request it). They also have pasture raised chickens, pork, and eggs from free-range laying hens. Going for the sirloin is a safe bet, but why not try the marrow bones or tongue? And for those of us brought up on liver-wurst sandwiches, their wurst is beyond compare.

Young's Longrange Farm

702 Woodbury Rd., Watertown; 203-233-9163; youngslongrangefarm.com

This fourth generation owned and family operated farm specializes in non-GMO meats, but offers you everything else you need, seven days a week, Monday through Friday 3 to 5 p.m. and Saturday and Sunday 10 a.m. to 5 p.m. at their farm store. The farm grows and harvests seasonal vegetables and herbs summer to fall, and brings you treats like Goatboy Soaps and Cows Around the Corner milk. The pasture-fed steers provide juicy tenderloins and hearty briskets, while their pork comes in brats, links, dogs, and bacon. They also have seasonal chickens and turkeys for sale. Everything in the store is locally made or produced, and you will always find the meat freezers full.

Farm Breweries and Cideries

Hops on the Hill

275 Dug Rd., South Glastonbury; hopsonthehillbrewery.com

This "farm to glass" brewery makes all their beers with hops and grains grown on their farm and other family-run farms within 30 miles. Owner Al Gondek's family has owned and farmed this land since the 1950s, growing broadleaf tobacco, pumpkins, and other crops. Gondek added hops to the mix and turned an old tobacco drying barn into the brewery and tasting room. There is limited indoor seating in the barn, but a large area of picnic tables, complete with fire pits and propane heaters. Food trucks set up here almost every open day, so check their page for the schedule, or bring your own picnic to enjoy with a Maddie Peach Ale, Rusty Triangle, or Mel-Mel Stout. “Our hope is that when you drink our beers you have the same ‘mmmm’ reaction as when you take a bite of fresh baked bread still warm from the oven,” says Gondek. They also have two wines, two types of hard cider, and plenty of non-alcoholic options. Open Thursday and Friday 4 to 8 p.m., and Saturday and Sunday 1 to 6 p.m.

Kent Falls Brewing Company

33 Camps Rd., Kent; 860-398-9645; kentfallsbrewing.com

Just down the road from the tumbling Kent Falls, this brewery is part of a 50-acre farm that has been in continuous agricultural use for 250 years. They raise poultry and pork, along with an acre of hops and acre and a half of cider apples. They work with other local farmers, use a solar thermal array, and compost all of the spent grain, yeast and hop trub, and spent fruits with wood chips to create rich natural fertilizer to re-invigorate their plants. You can find their beer across the state and beyond, but why not visit their tasting room to enjoy it? The tasting room is open Thursday and Friday from 2 to 7 p.m., and Saturday and Sunday 12 to 5 p.m. They also offer barnside pickup during those hours. Try their Danny Boy pilsner or our favorite, the Partage de peche, brewed with local malt, local raw wheat, and aged hops from the farm. This was the first farm brewery in Connecticut, pioneers of a trend that has taken root just in the last decade. Beer is an agricultural product, and it is about time we felt that connection more keenly.

Mine Hill Distillery

5 Mine Hill Rd., Roxbury; 860-210-1872; minehilldistillery.com

Mine Hill Distillery makes gin and vodka from locally sourced corn, rye, wheat, and malted barley. Owner Elliot Davis raised his children on a farm just down the road and is working with farmers the way agriculture is supposed to work—as partners in the craft. The spent grain from the process then goes to feed heritage livestock at the local farms, as well. The classic dry gin includes eastern white pine and the classic vodka is distilled from their own wheat base. The distillery and tasting room are open to the public 12 to 4 p.m. on weekends.

Norbrook Farm Brewery

204 Stillman Hill Rd., Colebrook; 860-909-1016; norbrookfarm.com

This incredible 450-acre farm in Colebrook is a great place to walk, snowshoe, mountain bike, or even play disc golf, with maps, trail signs, and more being added every day. But the real attraction here is the beer. They specialize in farmhouse-style beers, ales, and lagers, with their six signature beers always on tap, and more being crafted every day. The Mount Pisgah IPA is smooth and balanced, with a caramel note, and the Beckley Furnace Brown Ale is toasty and nutty, perfect for a winter's day after cross-country-skiing the

trails at the farm. They are open Wednesday through Saturday 12 to 8 p.m. and Sunday 12 to 6 p.m., and the trails close at sunset.

Farmers Markets

Goshen Farmers Market

42 North St., Goshen; 860-248-1082; goshenfarmersmarket.com
For over a decade, the Goshen Farmers Market has searched for a permanent home in this tiny community and has found one at Scoville Park at the town hall. The whole town pitches together for these weekly markets, and promotes local farmers, crafters, and artisans through them. Live music keeps you entertained Saturday 10 a.m. to 1 p.m. from July until September. There is also a winter market, so check their website for hours.

Litchfield Farmers Market

Woodruff Ln., Litchfield; litchfieldfarmersmarket.org

Starting in 2007, Litchfield has created a year-round market that brings visitors from all around the northwest part of the state. *Cooking Light*'s online magazine recognized this as one of the 50 best farmers markets in the country, and if you attend, you'll find out why. A program of Sustainable Healthy Communities, a public charity focused on promoting healthy eating and active lifestyles in the Northwest Hills, this market brings sustainable, local food to a central location where everyone can find baked goods, organic produce, local honey, artisanal cheeses, and a variety of crafts. The outdoor Farm-Fresh Market is open Saturday, 10 a.m. to 1 p.m. in the Center School parking lot, while the winter indoor market is held at the Litchfield Community Center, 421 Bantam Road, from mid-October to mid-June. "Local food is healthier food," say the folks at Farm-Fresh Market, and we should all take note of that wisdom.

New Milford Farmers Market

New Milford Town Green, New Milford; 860-350-0676; newmilfordfarmersmarket.com

This market brightens up the New Milford town green every Saturday from Mother's Day through October, 9 a.m. to 12 p.m., with a wide selection of farm goods, baked goods, and hand-crafted products. With tomatoes and flowers, corn and scones, this market brings local products to the community, and brings in people from miles around.

Norfolk Farmers Market

19 Maple Ave., Norfolk; 860-542-5044; norfolkfarmersmarket.org

The charming hamlet of Norfolk runs a farmers market 10 a.m. to 1 p.m. every Saturday, June through early September, with numerous local vendors, food assistance for local families, and cooking demos by chefs every other week. You can get all the farmers market goodies you would expect, from raw goat milk to grass-fed beef, organic vegetables to homemade granola. But you can also find flower arrangements, ornamental trees, hand-dyed yarn, painted glassware, and turned wood. This is a crafter's paradise amongst the farmers markets of the northwest.

West Hartford Farmers Market

Corner LaSalle Rd. and Arapahoe Rd., West Hartford; 860-982-0673; whfarmersmarket.com

The corner of LaSalle Road and Arapahoe Road has been graced by this farmers market since 1992. More amazingly, the core group of farmers who began this market is still there, providing local Connecticut food to the people of West Hartford. In normal years, the market schedules special programs and events that accompany the smell of Wave Hill Breads fresh sourdough and crunch of Newgate Farms fresh bok choy. They run from early June to December, Saturday 9 a.m. to 1 p.m. and late June through September, Tuesday 9 a.m. to 12 p.m. West Hartford is considered one of the best places to live in America, and this market is one of many reasons why.

Wethersfield Farmers Market

220 Hartford Ave., Wethersfield; 860-578-8650; wfmarket.org

On the beautiful expanse of the town green, within sight of the lapping waters of Wethersfield Cove, this market has grown a lot from its first year in 2008. Now a destination market, people often combine a trip here with a stop at the Webb-Deane-Stevens House and the other sights of the old town (tied with Windsor for the oldest settlement in the state). Everything at this market is CT Grown certified, and with over 30 vendors every week, that's a lot of local businesses you are helping. You'll find organic produce, herbs, flowers, cheese, honey, and more. Live music and one or two food trucks join the crowd, and you might even find a book truck, too! From candles to candy, popcorn to orchids, this market has it all, Thursday 3 to 6 p.m. starting in May and lasting far into autumn.

Flamig Farm

7 Shingle Mill Rd., West Simsbury; 860-658-5070; flamigfarm.com

If you see someone walking around Connecticut with a backwards EGGS t-shirt, this is where they got it. Founded in 1907, the small 10-acre Flamig Farm was once a thriving dairy operation, but then turned to egg retailing, with thousands of chickens. Their current incarnation as a petting zoo has made them legendary in the Farmington River valley, and there is seldom a child who has not visited to pet the goats sometime April through November from 9 a.m. to 5 p.m. It is also a popular place for Halloween hayrides and weddings. However, unknown to many of these summer visitors, it is also a farm stay, with a comfortable apartment located above the

farm store. You have a balcony with a grill and a view overlooking the animals and fields. With two bedrooms that include a queen-size bed and space for three children, this is the perfect family getaway that kids will go crazy for. Air-conditioned in the summer and heated by a pellet stove in the winter, with a stocked kitchen and smart television, this is the farm vacation you've always wanted.

Maple View Farm

192 Salmon Brook St., Granby; 860-655-2036; mapleviewhorsefarm.com

Maple View Farm could have fit in many categories in this guidebook. It is a 50-acre, fourth-generation family farm that raises beef, pork, and chickens and sells them at their charming farm store 8 a.m. to 8 p.m. every day. From late November to Christmas, they sell pre-cut Balsam fir trees. They could feature in the trail ride section because they offer Saturday pony rides for beginner riders aged 3–12. They have also recently built a farm brewery, with beer brewed with local ingredients accompanied by live music, food trucks, and their picnic menu (including their own delicious sausages!), open Friday 3 to 8 p.m., Saturday 1 to 8 p.m., and Sunday 1 to 6 p.m. However, what really puts the icing on the cake of this agritourism diversification is their update of the 1935 four-bedroom farmhouse, with full eat-in kitchen, two full bathrooms, and sleeping space for 9. You will wake to the rooster crowing and wonder why you ever lived in the city. This is a multiple attraction, but in some ways, Maple View Farm stands alone. That's one reason to keep it here amongst the rare Connecticut farm stays.

Festivals and Fairs

Berlin Fair

430 Beckley Rd., East Berlin; 860-828-0063; ctberlinfair.com

The Berlin Fair has been around in one manifestation or another since 1882, with the current version held by the Lions Club since 1948. They have all the usual country fair delights—animal pulls, crafting contests, and pie-eating contests. Dreamland Amusements provides a selection of rides and games of chance and skill, while live concerts keep you humming along the midway. But it might be some of the more unusual attractions that make the Berlin Fair so special. A nail-driving contest? A frog jumping contest? And though the traditional foods like the Berlin Congregational Church roast beef sandwich might be the most popular, a deep-fried Oreo might be the thing that keeps you coming back year after year to this unique fair. Held on a weekend in September, Friday 11 a.m. to 10 p.m., Saturday 9 a.m. to 10 p.m., and Sunday 9 a.m. to 7 p.m.

Bethlehem Fair

384 Main St. North, Bethlehem; 203-266-5350; bethlehemfair.com

Established in 1924, the Bethlehem Fair Society provides agricultural education and promotes rural living, every year the weekend after Labor Day. They feature over 200 (!) vendors of food, crafts, and entertainment at the fairgrounds. You can stuff your face with kielbasa and fajitas, clam cakes and funnel cakes, cotton candy and candy apples. You can shop for musical instruments or patio furniture, for pewter jewelry or motorcycles. And you can ride the pirate ship or a camel, take a chance at a game of chance, or take the 10-story Ferris wheel into the sky.

Connecticut Antique Machinery Association Fall Festival

1 Kent Cornwall Rd., Kent; 860-927-0050; ctamachinery.com

For some romantics, farming means a life without machines. This is not the view of the farmers themselves (except for maybe Eric's relatives in Pennsylvania Amish country). Farmers have almost always adopted the latest technology to help them more efficiently manage their land. The annual festival at the Connecticut Antique Machinery Museum, usually a quiet little collection of history museums in some restored buildings in Kent, is a place to find some of these amazing machines, as well as other antique machines of our American past. Along with tractors and agricultural machines of various sorts, you will find old trains, engines, and mining machines, all chugging away to the great delight of all attending. The festival also features wood splitting, blacksmithing, broom making, and cider making. Antique cars and trucks fill the fields, and food is available from a variety of vendors. Also on-site are the Cream Hill Agricultural School, with its original desks and library, and the Connecticut Museum of Mining and Mineral Science, full of native rocks and mining equipment. Literally next door is the separately run Sloane-Stanley Museum, featuring a wonderful collection of colonial American tools and the partially restored ruins of the Kent Iron Furnace.

Connecticut Wine Festival

116 Old Middle St., Goshen; 860-216-6439; ctwine.com/events/wine-festival

Every year on a hot summer day at the Goshen Fairgrounds, the Connecticut Wine Trail holds its annual festival. There is live music, crafts, and food. But the reason to come and browse the farm buildings is that dozens of local winemakers have gathered to offer you a taste of their wines. You will be overwhelmed with the choices of reds and whites and rosés and sweet dessert wines, befuddled by the hints of chocolate and cherries and spice. A day spent here will be an education for your palate, and a memory for years to come. One caution: be sure to bring a designated driver to this event!

Four Town Fair

56 Egypt Rd., Somers; 860-749-6527; fourtownfair.com

During the winter of 1838–1839, when the farmers of Somers yoked 210 pairs of cattle for an exhibition, the Four Town was born. Due to incredible interest, the following October, the newly formed society held its first regular cattle show, and soon the towns of Somers, Enfield, Ellington, and East Windsor were holding an annual event, slowly adding pigs and chickens, crafts and contests. Today, you can participate in or just watch contests for the best livestock, vegetables, canning, needlework, crafts, flowers, art, baking, photography, and the ever-popular women's skillet throw. Can you eat more pie than the next person? What about corn? Do you boast the heaviest pumpkin? Come for the food and stay for the music every September for four days, Thursday 4 to 10 p.m., Friday 4 to 11 p.m., Saturday 8 a.m. to 11 p.m., and Sunday 8 a.m. to 6 p.m.

Goshen Fair

116 Old Middle St., Goshen; 860-491-3655; goshenfair.org

Held Labor Day weekend every year, the Goshen Fair is fun for the whole family, with live music, magic shows, fried food, and more. But amid the circus of the midway, it is always good to remember that this is first and foremost an agricultural fair. Farm animals are the biggest reminder, with show quality animals abounding, including dairy cattle, beef cattle, llamas, dairy goats, sheep, swine, poultry, and rabbits and cavies. You will be astonished by the size of blue-ribbon vegetables and fruit and the elaborate intricacy of needle-work and sewing. Some take the opportunity for education and learn how to raise a farm animal or forge a horseshoe from experts in the fields. Other pick a favorite antique tractor and cheer it on during the Pull and Display in the Motor Sports Arena.

Harwinton Fair

150 Locust Rd., Harwinton; 860-485-0464; harwintonfair.com

Connecticut has many town fairs, but not many that have been going steadily since 1853. Appearing early each October for three days Friday through Sunday, the Harwinton Fair is one of the largest fairs as well, with hundreds of attractions, exhibits, vendors, and more. There are displays, like a blacksmith shop and working shingle mill; demonstrations of pioneer and Civil War living; and competitions, such as baking and wood chopping. The fair hosts a large traveling carnival with rides, a high-wire act by circus performers, and dozens and dozens of craft and food vendors. They have entertainment for kids, like magic shows, and music entertainment for adults. Skillet tossing, pig racing, oxen teamster challenges, and an antique tractor show . . . what more do you want from a country fair that has been going for more than 160 years?

Shad Derby Festival

261 Broad St., Windsor; 860-688-5165; windsorshadderby.org

Derby Day in Windsor is the high point of two weeks of festivities celebrating the arrival and harvest of this unusual fish. The shad is a bony (though plump) relation of the herring, which migrates north along the Connecticut River to spawn, and Windsor is one of their destinations. Sign up to fish the river, or just come for the shad and festivities. There is a prize for the largest shad caught, and other events that vary, like a road race or a golf tournament. This is a townwide festival, and Windsor's Broad Street is usually packed with people and booths. If you get a chance to try these seasonal treats for dinner, don't miss it; for the more adventurous, shad roe fried with bacon is a delicacy. Events surrounding the festival begin in March or April, but early May is when they usually have the fishing tournament. In that way, the fish are in charge, since their decision when to migrate determines the dates of the festival.

Southington Apple Harvest Festival

Town Green, Southington; 860-276-8461; southington.org

Every October since 1968, 100,000 people have descended on Southington for this annual festival. It lasts a week, beginning with a parade and proceeding through arts and crafts shows, a carnival, a road race, a talent show, fireworks, music, and dancing. Of course, the mighty apple is featured heavily, with all sorts of apple-based foods to eat (as well as other delicacies like fried dough and lobster pies). You can try your luck at apple pie–eating contests, bed races, and singing competitions. The nearby orchards get into the fun, with wagon rides and apple picking. Parking and admission are free, and so is the live music and entertainment, centered around the town green but spilling into the streets and parks of Southington.

Maple Sugarhouses and Apiaries

The Humblebee Honey Company

20 Main St., Oakville; 203-565-5057; humblebeehoneycompany.com

Established in 2010, the Humble Bee Honey Company has several apiaries around the Litchfield Hills and is a member of the Connecticut Farm Bureau, The Backyard Beekeeper's Association, The Eastern Apicultural Society, and The American Beekeeping Federation. They design their apiaries amid flowering gardens for their health and livelihood, and produce and bottle honey and raw pollen, as well as make a skincare product line. They sell their products in grocery stores and at other farm stores, but also at their recently opened "farm store" in Oakville. It doesn't seem too farm-like at first glance, but amazingly has an eight-frame observation hive. You can also get other local products there, as well as tour their "honey house," where they extract and bottle the liquid gold. They farm their livestock (yes, honeybees are considered livestock) Monday through Thursday, and have hours at their store Wednesday and Thursday 4 to 7 p.m., Friday 1 to 6 p.m., Saturday 1:30 to 4 p.m., and Sunday 11:30 a.m. to 4:30 p.m. If you are really serious, they also give lectures on the gentle art of beekeeping to potential keepers and amateur gardeners alike.

Lamothe's Sugar House

89 Stone Rd., Burlington; 860-675-5043; lamothesugarhouse.com

The Lamothe farm is not old by Connecticut standards, starting in 1971. However, they quickly expanded to have 5,500 tapped maple trees, 15 miles of tubing, and a state-of-the-art sugarhouse, which you can tour seasonally from mid-February through March, weekends from 1 to 4 p.m. Their country gift store, open year-round, is the premier place in Connecticut to get maple syrup and maple products (other Connecticut-made items, even farm-raised pigs, can be ordered). My favorites are the maple sugar–coated nuts, but others swear by the maple saltwater taffy. "Grade A Dark Amber or nothing" is what hard-core maple lovers say. If you take your coffee with sugar, try their own Maple Farmhouse Blend with a little of Lamothe's maple syrup instead. You'll never go back.

Sullivan Farm

140 Park Lane Rd., New Milford; 860-210-2030; sullivanfarmnm.org

Sullivan Farm is a vocational and educational agricultural center that benefits local youth by giving them hands-on skills in farming and agribusiness. From late February through the end of March, the staff and students tap over 1,600 trees to produce maple syrup, one of the largest operations in the area. The Great Brook Sugar House on the farm is open to the public during those months, and you can call or "watch for the steam coming from the cupola." They hold an open house on a Saturday toward the end of the season to teach sugaring history and the modern operation. Their farm stand is also operated by the youth staff, and open from the third week of May until Christmas Eve, Sunday 12 to 5:30 p.m., and every other day 10 a.m. to 5:30 p.m. You can buy local produce that changes with the seasons, as well as maple syrup, of course.

Sweet Wind Farm

339 South Rd. (Rte. 179), East Hartland; 860-653-2038; sweetwindfarm.net

Operated by Susan and Arlow Case Jr., Sweet Wind makes delicious syrup at Arlow's Sugar Shack. Tours and classes are offered during mapling season (February and March) and Saturdays are open houses. The family often runs a maple festival in March with free educational events, along with gardening lectures and workshops, so check their website for event times and sign up. They have a CSA farm share/box either by the season or weekly, including their maple syrup, and also fresh fruit, vegetables, herbs, jelly, firewood, and flowers. Their farm store is open year-round, daily 10 a.m. to 6 p.m., and offers the farm's own chemical- and pesticide-free strawberries, blueberries, raspberries, tomatoes, pumpkins, squash, and more.

Museums and Education

4-H Auer Farm

158 Auer Farm Rd., Bloomfield; 860-242-7144; auerfarm.org

This 120-acre nonprofit educational farm under the shadow of Talcott Mountain may only be 2 miles from Hartford, but when your children are having fun and learning on the trails, orchards, and pastures, it feels a world away. Along with hosting school field trips and community events, Auer Farm has a science-based curriculum to challenge any child (or adult). They also feature goods from the farm itself, including eggs, maple syrup, honey, alpaca yarn (yes, they have alpacas), Christmas trees, blueberries, and farm-raised pork and beef. Usually open Monday through Saturday 8 a.m. to 5 p.m., but check their current hours and call ahead to make an appointment for an experience you and your children won't soon forget.

Beavertides Farm

44 Cobble Rd, Falls Village; 860-824-0421; beavertidesfarm.com

This small farm run by Marleen and Dan provides grass-fed sheep and goat meat, along with honey and bee-related products, to the local community. You can get them by joining their Farm Club and pre-ordering boxes of their meat online. They also have a small cabin farm stay, where you can cook products from the farm on an old-fashioned woodstove. But the real draw at Beavertides might be their educational programs, focusing on regenerative agriculture, animal husbandry, and sustainable living. The popular beekeeping training course is the standout. The course usually runs on Saturday mornings beginning in April, all the way to November. It is the perfect way to get started with this vital practice of keeping bees to not only make great honey but to help pollinate the fruit and vegetables we all need to sustain this planet.

Burlington Fish Hatchery

34 Belden Rd., Burlington; 860-673-2340; portal.ct.gov/DEEP/Fishing/Fisheries-Management/Burlington-Hatchery

Fish farming has been going on for thousands of years and has taken leaps and bounds recently, especially in Asia. However, it has made inroads (or streams?) into places like Connecticut. Aquaculture is happening in Long Island Sound, of course, but also in private hatcheries in places like Newtown and Waterbury, some of it for stocking lakes and streams so that we can harvest fish ourselves, and some of it for sale to restaurants. The Burlington Fish Hatchery, run by the state, has been doing it for almost a century, developing brown trout, brook trout, kokanee salmon, and rainbow trout to stock our rivers and lakes and eventually make it onto our kitchen tables. They are open to visitors every day from 8 a.m. to 3:30 p.m., and you can walk around the outdoor ponds of fingerlings and the indoor spawning pools. This brief education will help you understand the world of aquaculture, as we begin to contend with the food needs of a growing population.

URTESY OF TRENA LEHMAN

Comstock Ferre and Company

263 Main St., Wethersfield; 860-571-6590; comstockferre.com

During the 19th century Wethersfield made its reputation by selling seeds. Since 1811 seeds have been sold from this spot on Main Street, and since 1838 the business has operated under the name Comstock Ferre and Company, the longest continually operating company of its kind in the US. At their "heirloom market" you can find open-pollinated, pure, non-GMO seeds of the highest quality, heirlooms that you won't find anywhere else. These are herbs, vegetables, and flowers in primal forms—healthier and with different nutrients—that you can grow at home. You can also get all sorts of plants, natural food products, handcrafts, and antiques. It's a fun store and a living museum all in one.

Connecticut Valley Tobacco Museum

135 Lang Rd., Windsor; 860-285-1888; tobaccohistsoc.org

The tobacco here in the Connecticut valley is primarily used as cigar wrappers, and it has been used to wrap the world's finest. The Luddy/Taylor Connecticut Valley Tobacco Museum includes a restored tobacco-curing barn, where authentic equipment shows the process of harvest and curing. Historical displays in the museum next door show the industry's contribution to the region's economy and culture. There are live animal exhibits in the animal barn and nature center, and a concert series is held from September to May. The 475 acres surrounding the museum, called Northwest Park, include 12 miles of trails through a variety of habitats that are great for mountain biking or for cross-country skiing in winter (northwest park.org). They also have an annual cigar barbecue for all those smokers out there. Open Thursday and Friday 12 to 4 p.m. and Saturday 10 a.m. to 4 p.m.

Flanders Nature Center and Land Trust Sugar House

5 Church Hill Rd., Woodbury; 203-263-3711; flandersnaturecenter.org

Started by Natalie Van Vleck, who raised turkeys and sheep in the early 20th century, the Flanders Nature Center and Land Trust own an incredible number of acres in Woodbury, preserving forest and farmland for future generations. They have special events like farm-to-dinners and run programs for schools and for smaller groups throughout the summer, teaching land stewardship and appreciation for nature and art. However, many people outside northwest Connecticut also know the Flanders Nature Center for the maple syruping demonstrations in February and March. Call to pre-register for demonstrations Saturday and Sunday from 1 to 4 p.m. and, of course, pick up some of the syrup while you are there.

Toplands Farm

102 Painter Hill Rd., Roxbury; 860-354-0649; toplandsfarm.com

Since 1942 the Diebold family has operated this large farm, raising heritage breed hogs, chickens, and beef cattle. They locally source grains to feed their cattle and raise their own hay. "By raising small groups of animals, we are able to provide them with plenty of room to move around, a more complete diet, and plenty of water," they say. Their farm stand is open Monday through Friday during the afternoon and Saturday 10 a.m. to 4 p.m. However, they also feature the Double D Living History Farm, which is the largest collection of antique farm equipment in the entire northeast. Take a tour and watch demonstrations of the tractors and engines, April 1 to

November 1, by appointment. This place is an education in how agriculture once worked, and how country life continues to this day.

Pick Your Own

Easy Pickin's Farm

46 Bailey Rd., Enfield; 860-763-3276; easypickinsorchard.com

This former tobacco farm had a small orchard of two dozen apple trees planted in the 1930s, and in the 1960s Francis ("Red") and Linda Kelliher began to plant more. In the 1970s and 1980s their sons planted even more, along with raspberries and peaches. In 1992 Easy Pickin's became the pick-your-own star of Enfield, and since then has added more and more options, from flowers and herbs to plums and Asian pears. They also have a CSA to sign up for and bring their delicious produce to many local farmers markets. In the autumn there are free wagon rides on Sunday afternoons on the hour, along with make-your-own-scarecrow activities for the kids. One unique offering is the annual gourd hunt, a tradition since 1995. Think of it as a fall Easter egg hunt, except instead of small eggs, the kids hunt for these decorative prizes in the orchard. Check their website or Facebook for current hours (often 9 a.m. to 5 p.m. during the picking season).

Ellsworth Hill Orchard & Berry Farm

461 Cornwall Bridge Rd. (Rte. 4), Sharon; 860-364-0025; ellsworthfarm.com

Pick your own strawberries, blueberries, raspberries, flowers, and more at this gorgeous spot high on the ridge above the Housatonic River Valley. The farmers here call the land "Mother Nature's Candy Store," and no wonder. The pick-your-own stars may be the apples, but they also have strawberries, cherries, raspberries, plums, blueberries, peaches, pears, and pumpkins. Ellsworth Hill also has vegetables, flowers, and harvested fruit at their retail store, open every day but Tuesday, 9 a.m. to 6 p.m., and seven days a week for pick-your-own June to November. Head to Ellsworth early in the apple season on September 16 or so for the Honeycrisps and Macouns or wait for October 20 and harvest the Northern Spy Idareds. If you're there between Labor Day weekend and November, try the fall corn maze. But wear sturdy shoes and take a compass, because your kids might intentionally try to get lost so they can stay on the farm. It's that much fun.

Harris Hill Farm

106 Ridge Rd., New Milford; 860-354-5856; harrishillfarm.com

This fascinating farm opens its gates on October weekends every year for a pick-your-own pumpkin patch. For more than 20 years, people have been coming to get their jack-o'-lantern pumpkins here, along with gourds and edible squashes. In the past they have featured hayrides and farm animal visits, so check their website. The barn at Harris Hill is also part of the New Milford Barn Quilt Trail (newmilfordfarmlandpres.org), with a colorful brown cow that honors their history with Brown Swiss cattle. Barn quilts are another way that agricultural tourism is once again becoming a community activity that creates local pride of place. Let's hope more towns take a page from New Milford's book, or rather, a square from their quilt.

Litchfield Hills Blueberry Farm

23 Schrowback Rd., Plymouth; 860-283-9571; litchberry.com

The Litchfield Hills Blueberry Farm has been family owned and operated since 1998, though the blueberry crops pre-date the current owners. Twelve thousand highbush blueberry bushes spread over a 30-acre hillside, with many different varieties. What's that? You are not familiar with Blueray or Bluetta? What about Herberts or Wolcotts? They have some available early, some late, some in-between, pruning the bushes for reachable picking. The fields are open Tuesday to Friday, 8 a.m. to 8 p.m. and Saturday and Sunday 9 a.m. to 4 p.m., while supplies last in July and August.

Lost Acres Orchard & Farm Store

130 Lost Acres Rd., North Granby; 860-653-6600; lostacres.com

Not to be confused with the vineyard just down the road, this small family farm along Belden Brook has a bakery and farm store and posts menus online daily for their fans and customers. It is also a great place to pick apples and peaches during the summer and early fall. They are usually open Wednesday through Sunday from 9 a.m. to 5 p.m. This is close to the most rural area of the state, so keep traveling west along Route 20 and find the sort of idyllic peace you thought wasn't available in Connecticut.

Maple Bank Farm

57 Church St., Roxbury; 860-354-1278; maplebank.farm

This family farm dates back to 1730, but the newest Maple Bank farmers are Dakota, with a degree in sustainable food systems from Sterling College and expertise in knife-making and pottery, and Matthew, who studied philosophy and physics and worked as an outdoor educator leading wilderness and whitewater kayaking trips. But both dreamed of owning a farm and now act as stewards of Maple Bank, with a farm stand open Tuesday to Sunday 10 a.m. to 5:30 p.m. where you can buy dairy, meat, coffee, fresh bread, and more. Their pick-your-own blueberries are the real draw here, though, Wednesday to Sunday 8 a.m. to 5 p.m. in July and August. It's important to support young farmers like Matt and Dakota, because without them, the traditions and lore will die. And that would not be good news for our bellies or our world.

March Farm

160 Munger Lane, Bethlehem; 203-266-7721; marchfarm.com

This fourth-generation family farm has pick-your-own opportunities for apples, blueberries, cherries, nectarines, peaches, plums, pumpkins, and strawberries. At their farm store and bakery, you'll find veggies and eggs and bread and all the good local things you'd expect. But March Farm is even more—with live concerts, farm-to-table events, a fruit harvest festival, and a 30,000-square-foot farm playscape and 5-acre corn maze that keeps kids entertained for hours. This is destination agritourism at its best. It's open seven days a week May through December, 10 a.m. to 6 p.m. The pick-your-own season starts in mid-June with cherries, then transitions to blueberries and peaches in mid-July, and apples and pumpkins in September and October. If you're in the area, they also have a 16-week CSA June through September.

The Pickin' Patch

Nod Rd., Avon; 860-677-9552; thepickinpatch.com

Since 1666 this land has been a family farm, and the current descendants, Janet and Don Carville, preside over one of the most popular pick-your-own farms in the state. The strawberry patch is particularly large and famous, though you can also come here for blueberries, Christmas trees, vegetables, and a variety of other things. From April until December, you can visit from 10 a.m. to 5 p.m. daily. October weekends they have hayrides with the Pumpkin Lady to the patch. Buying directly from a farm is often cheaper than retail, and it's a head scratcher why more people don't take advantage of it.

Rogers Orchards

336 Long Bottom Rd., Southington; 860-229-4240; rogersorchards.com

Since 1809 this orchard has been a Connecticut landmark, with eight generations of farmers behind it. It was founded in 1809 by Chauncey Merriman, a veteran of the Revolutionary War, and today his great, great, great, great grandson, John Rogers and his family manage the farm. In 1965 Rogers opened a second location at 2876 Meriden Waterbury Turnpike in Southington, its own 90-acre farm with a separate farm store (and phone, 203-879-1206). Both stores sell their fruit, vegetables, honey, maple syrup, jams, jellies, and preserves. Their bakery makes flaky 10-inch pies (try the pecan!), apple cider donuts, apple sour cream coffee cakes, and cookies. They are

open seven days a week from late July through mid-May and offer pick-your-own apples on weekends in September and November. The orchard has an astonishing 20 varieties of apples, and also grows peaches, nectarines, plums, apricots, and pears.

Rose's Berry Farm

295 Matson Hill Rd., South Glastonbury; 860-633-7467; rosesberryfarm.com

With two locations, Rose's has twice the opportunities for you. The Matson Hill location is the picking farm, with blueberries, raspberries, and strawberries throughout the summer. The farm stand is at 1200 Hebron Avenue in Glastonbury, with fresh produce, pies, hanging plants, herbs, fresh-cut flowers, and a wide variety of vegetables, from snap peas to summer squash. Rose's also offer fall hayrides on Saturday and Sunday during the fall and delicious Sunday morning farm breakfasts starting in early June and ending in early October, 8 a.m. to 1 p.m. If you stop at the farm stand, don't forget to pick up some of their homemade jam.

COURTESY OF TRENA LEHMAN

Scantic Valley Farm

327 Ninth District Rd., Somers; 860-749-3286; scanticvalleyfarm.com

Scantic Valley Farm has hill-top strawberry fields for picking, beginning in June. Their exquisite blueberries become available in mid-July. At their stand, they also sell belted Galloway Beef and heritage pork, along with jam, honey, maple syrup, and field-fresh flowers. Scantic Valley is open 8 a.m. to 6 p.m. weekdays and 8 a.m. to 2 p.m. weekends. Remember, as with all your picking opportunities, those hours might be weather dependent, so call ahead. It's a good idea to check what's ripe anyway!

Trail Rides

Bradley Mountain Farm

537 Shuttle Meadow Rd., Southington; 860-385-GOAT; bradleymountainfarm.com

This 200-year-old dairy farm on Crescent Lake was built by Ichabod Bradley in 1813 and is on the registry of National Historical Places. Their goat-milk soap is the prime product here, made from the herd of LaMancha, Mini-LaMancha, and Nigerian Dwarf goats. They also run tours of the historic Bradley Home, with its 9 fireplaces and 7 authentic colonial murals. But the reason this farm is so beloved is its goat-related activities. Kids of all ages (including middle and old ages) can schedule play dates to groom, cuddle, and walk your goat on a lead on the property. Goat strolls and hikes in the pastures are a popular form of relaxation, along with the slightly sillier goat yoga. They have more hands-on instruction, too, as you accompany the farmers to take care of the herd and other farm animals.

Country Quilt Llama Farm

P.O. Box 21, Cornwall; 860-248-0355; countryquiltllamafarm.com

Llamas living in Connecticut might seem odd to some, but Debbie Labbe knows that they are much more than curiosities. They are perfect animal ambassadors, well suited to interact with people.

Debbie started the farm, located in West Cornwall, in 1988 taking her two llamas on the road to meet preschoolers and seniors. Those special events are still available, and Debbie holds "Hooked on Llamas" education and animal therapy programs for young and old. But there's more. Llama walks are offered in partnership with White Memorial Conservation Center in Litchfield. You'll have the chance to stroll around the 35 miles of trails and get to know Snoopy, Harley, Rio, Robert, Bentley, Cooper, Jack, and Theo. These guys are perfect companions to walk in snow and sun, so call Debbie to schedule your trek. You'll come away with deeper connection to these animals and the path you travel together.

Lee's Riding Stable

57 E. Litchfield Rd., Litchfield; 860-567-0785; windfieldmorganfarm.com

The Windfield Morgan Farm, set on 100 beautiful acres adjacent to Topsmead State Forest in Litchfield, has been raising primarily Morgan horses (thus the name) since 1976. Many show champions have come from here, and their breeding program is well respected. But unless you're looking to buy a horse, the adjunct Lee's Riding Stable at the farm is probably what you're interested in. The friendly horses are perfect for riding in the rolling hills of Litchfield, and guided riding tours are available 9 a.m. to 5 p.m. seven days a week. Call and make a reservation for up to eight riders. Lee's also offers a variety of lessons for riders of all ages, including pony rides for very young horse lovers. The people and horses here are some of the friendliest you'll find in the state.

Rustling Wind Stables

164 Canaan Mountain Rd., Falls Village; 860-824-7634; rustlingwind.com

This 200-acre farm has been in the Lamothe family since the 1800s, and today the fourth generation is on the land. Since 1965 Rustling Wind has been a horse farm and today has 30 horses, an indoor and outdoor arena, a jumping field, and miles of trails. If you are 8 or older, you can take guided English or Western saddle rides for one to two hours through the forests of this picturesque area, and they do have pony rides for smaller children. Their summer hours are usually Tuesday through Saturday 9 a.m. to 12 p.m. and/or 1 to 4 p.m. The farm is also a creamery, and at their store, open 8:30 a.m. to 4:30 p.m. daily, they offer 10 different delectable cheeses, as well as jams, pickles, maple products, goat's milk soap, and hand knits from their own sheep wool. When you're done exploring the trails on horseback and sampling the cheese, be sure to stop at the Great Falls of the Housatonic, for which the village is named.

Tree Farms

Burgess Nursery

373 Deming St., South Windsor; 860-815-2333; burgessnurseryct.com

Owner Larry W. Burgess and his son Patrick offer seasonal produce using organic and sustainable practices. They specialize in garlic and heirloom tomatoes, which you can find at their farm stand, Saturday and Sunday 9 a.m. to 5 p.m. But as a member of the Connecticut Christmas Tree Growers Association, they also start selling trees beginning the day after Thanksgiving every year. This is a true small family farm, and those are often the ones doing the hardest and best work.

Busy Acres Tree Farm

548 Quassapaug Rd., Woodbury; 203-695-4785; busyacrestreefarms.com

Since 1979, Busy Acres has been growing Christmas trees, with 30 acres of cut-your-own Fraser Fir, White Pine, White Spruce, Blue Spruce, Norway Spruce, and Douglas Fir. Along with the normal-size selection of trees, they also have some 12 feet or taller. Of course, you should make sure that you have space in your living room (and your vehicle) before buying one! The farm opens the day after Thanksgiving 9 a.m. to 3:30 p.m. and is open after that Tuesday through Friday 12 to 3:30 p.m. Dress warm, bring cash or checks, and a camera for those photos of the whole family that everyone loves to receive with their annual cards.

Deeply Rooted Farms

Rte. 72, Harwinton; 860-921-3434; deeplyrootedfarms.net

Deeply Rooted Farms began in 1973 with a family Christmas tree stand and has since expanded to include a pick-your-own strawberry farm in Harwinton. You can visit the 5 acres of strawberry picking fields throughout June and July (and sometimes earlier or later), so check the "news from the field" section of their website for updates. They also have pre-picked berries for sale at their stand and at local markets. And they raise curcubits. What are those, you ask? Well, that's a catch-all term for specialty gourds, pumpkins, and winter squash, some of the first edible cultivated plants in both the Old and New Worlds. You can find those at local markets and other farms in the area. Deeply Rooted also continues its tradition as a tree farm, with multiple fields of Balsam Fir, Canaan Fir, Concolor Fir, Fraser Fir, White Spruce, Blue Spruce, Meyer Spruce, and White Pine. Check in at 91 Terryville Road before harvesting, the day after Thanksgiving through December 24; weekends 9 a.m. to 5 p.m. and Friday 12 to 5 p.m.

Dzen Farms

187 Windsorville Rd., Ellington; 860-871-8183; dzenfarms.com

Dzen Farms was founded in the 1930s, originally raising potatoes, tobacco, and dairy cows. Although they have completely shifted from those original products, they smartly maintain a diverse farm, beginning to raise strawberries and Christmas trees in the 1970s and blueberries around 1980. The garden market (at the address above) opened in 1998 and carries fresh produce, flowers, native honey, and ice cream, with 24 flavors, over 10 toppings, milkshakes, nor'easters, and sundaes throughout the spring and summer. Open seven days a week, with the greenhouse hours from 8 a.m. to 8 p.m., and the ice cream shoppe open 12 to 8 p.m. They also have pick-your-own blueberries and strawberries in the summer at 87 Barber Hill Road, Broad Brook, 8 a.m. to 12 p.m. Monday through Friday and 8 a.m. to 4 p.m. Saturday and Sunday (cash or check only). The Christmas tree farm is actually at three locations: 575 Barber Hill Road, South Windsor; 174 Barber Hill Road, South Windsor; and 18 Sadds Mill Road, Ellington, for a combination of over 100 acres.

Walk into any one of these fields of trees and smell that green, piney scent. It smells like home.

Humming Grove Christmas Tree Farm

73 Barber Hill Rd., Broad Brook; 860-208-3517; humminggrove.com

Humming Grove sells premium wreaths, honey, beeswax candles, fruits, vegetables, handmade ornaments, and even gourmet mushrooms. Whew! That's a lot. But you're probably more interested in their trees, and although Humming Grove is not the largest tree farm in Connecticut, they certainly have a variety of species and sizes for any tabletop, living room, or even for the lobby of your corporate skyscraper, with heights up to 22 feet. Tagging starts on weekends following Halloween and the tree harvest starts, as usual, the Friday after Thanksgiving, Friday and Saturday 9 a.m. to 4 p.m. and Sunday 1 to 4 p.m.

Syme Family Farm

72 Windsorville Rd., Broad Brook; 860-623-5925; symefamilyfarm.com

This family farm in Broad Brook (part of East Windsor for those of us not in the area) specializes in flowers, with popular on-farm bouquet making workshops and farm tours every summer season. They sell these flowers, along with other herbs and potted plants, as well as farm fresh eggs and occasional other products during the summer and fall. And then, in November, the Christmas tree farm part of the operation gears up, and the Syme family welcomes you to the farm to cut your own trees and grab a wreath or two for the holiday season.

Wineries

Bethlehem Vineyard

46 Town Line Rd., Bethlehem; 203-266-5024;
bethlehemvineyard.com

Jim Riordan planted his first Cayuga and St. Croix vines in 2005 and opened to the public in 2013. Riordan has embraced Bethlehem's status as a "Christmas town" with wines like Old St. Nicholas, which is made from estate grown Traminette grapes. Their estate grown Cayuga is soft and sweet, like a German Riesling. In addition to growing grapes, Riordan also makes apple cider from a local orchard. The tasting room is open 1 to 5 p.m. on selected dates throughout the season, May to December.

Brignole Vineyards

103 Hartford Ave., East Granby; 860-653-9463;
brignolevineyards.com

Set in the heart of rural East Granby, Brignole Vineyards is a 15-acre, family-owned winery that makes wine like California, but looks perfectly at home in northern Connecticut. They feature both estate-bottled charmers like Frontenac and California-grape-based wines like Cabernet Sauvignon, for any palate. They also serve sangrias and slushees seasonally, great ways to cool off or warm up. Open Monday through Friday 1 to 7 p.m., Saturday 12 to 7 p.m., and Sunday 12 to 6 p.m.

Chateau Le' Gari'

303 South Main St., Marlborough; 860-467-6296;
chateaulegari.com

If Marlborough seems to you like it should be in Tolland or New London County instead of Hartford, you are not alone. But it is, and it is where Gary Crump is making wine these days, after decades of making classic, award winning wines elsewhere. He focuses on small-batch, estate-grown wines like Fawn Brook White, taking advantage of the gravelly soils and temperate weather of his micro-climate. The tasting room is open Friday and Saturday 11 a.m. to 6 p.m., and Sunday 12 to 6 p.m.

Connecticut Valley Winery

1480 Litchfield Turnpike. (Rte. 202), New Hartford; 860-489-9463; ctvalleywinery.com

This is a perfect short stop on Route 202 between Hartford and Torrington. Standouts at Connecticut Valley Winery include Deep Purple, made with Chambourcin grapes, and Midnight, featuring Frontenac. Both are estate-bottled, as is Dolce Vita, a lovely sweet white. The Italian heritage of the winemaking Ferraro family is evident in their Chianti and Spumonte Muscato, a sparkling white. Specialty wines include Black Bear dessert wine, named for a bear that helped itself to the harvest and found the grapes as inviting as you will. The tasting room's wood ambience and hearty fireplace offer respite on a late autumn day, and the patio umbrellas will shield your glass from the sun if you're out in the summer heat. Their normal open hours run from mid-February through mid-December, Saturday and Sunday 12 to 5 p.m.

Hawk Ridge Winery

28 Plungis Rd., Watertown; 860-274-7440; hawkridgewinery.net

Part of Hidden Breeze Farm, Hawk Ridge is named for the red-tailed hawks that hover over the 58 acres of the historic horse and hay farm. The tasting room is open every day, Wednesday to Saturday 12 to 8 p.m. and Sunday to Tuesday 12 to 6 p.m. Unlike many wineries, Hawk Ridge serves its own food, with charcuterie boards, chicken tenders, and truffle fries to accompany the crisp lemongrass flavor of Mohawk Sauvignon Blanc. They also partner with Woodbury Brewing Company and offer some of their beers on draft. With live music, a 1,200-square-foot deck, and a fancy private events room, this is agritourism at its most luxurious.

Hopkins Vineyard

23 Hopkins Rd., New Preston; 860-868-7954; hopkinsvineyard.com

Developed on land that has been farmed for over 200 years, Hopkins Vineyard perches on the hills just above Lake Waramaug. Visit the winery's 19th-century red barn, where you'll find the tasting room, wine loft, and gift store. Tour the facilities and vines and witness the pride of the winemakers and the beauty of the land. Standouts include estate bottled Chardonnay; Duet, a blend of Chardonnay and Vidal Blanc; and Lady Rosé, a lovely dry rosé with tobacco on the nose and strawberry notes on the palate. Hopkins's Cabernet Franc may be the best in the state, in an area known for this rich Old-World grape. Try Red Barn Red, a dry red with black currant overtones, and Sachem's Picnic, a red blend, served cool, which is a little sweeter and perfect for summer barbecues. Hopkins also offers peach wine, apple cider, a dessert wine, and an exquisite ice wine, made from Vidal Blanc picked when the grapes are frozen. Estate-picked Pinot Noir and Chardonnay are blended for the sparkling wine, made in the champagne style. Hopkins is open May through October, Monday through Friday 11 a.m. to 5 p.m., Saturday 11 a.m. to 7 p.m., and Sunday 11 a.m. to 6 p.m., as well as November through December, Sunday to Friday 11 a.m. to 5 p.m. and Saturday 11 a.m. to 6 p.m. Unlike some wineries (and most farms), Hopkins is also open for tastings in the winter: visit January through April, Friday to Sunday 11 a.m. to 5 p.m.

Jerram Winery

535 Town Hill Rd., New Hartford; 860-379-8749; jerramwinery.com

The tasting room at Jerram Winery is small, located down the gravel drive between two barns. But inside, the yellow brightness is inviting, especially on spring or summer days when the light hits just right. Jerram grows several varieties, including Marechal Foch, Vignoles, and Seyval Blanc. The wines offer tasters the chance to try these and sample unique blends, including S'il Vous Plait, made with Cabernet Franc grapes, and Vespers, a late-harvest dessert wine. You'll also find Aurora, the goddess of dawn, adorning the frosted bottle of the wine (and the grape) that takes her name. This makes a great stop if you're up seeing the sights in Riverton or hiking Peoples State Forest. Open May through December, Friday to Sunday, 11 a.m. to 5 p.m.

Land of Nod Winery

99 Lower Rd., East Canaan; 860-824-5225; landofnodwinery.com

History is rich at Land of Nod Winery. It is a nationally recognized bicentennial farm; that is nine generations of loving the land. Named for a Robert Louis Stevenson poem, Land of Nod offers an intriguing selection of fruit wines made from berries grown on-site. Their offerings include raspberry wine, a blend of blueberry and raspberry, peach wine, and a delicious chocolate-raspberry dessert wine. They also produce grape wines, including Bianca, a lovely white, and Marquette, a fruit-forward red. Tastings take place in a small shop that also offers yarn from the farm's herd of sheep, along with local maple syrup, gourmet snacks, and other gifts. Open seasonally April to May, Saturday and Sunday 11 a.m. to 5 p.m.; June through August, Friday through Sunday 11 a.m. to 5 p.m.; and September through mid-November, Saturday and Sunday 11 a.m. to 5 p.m.

Lost Acres Vineyard

80 Lost Acres Rd., North Granby; 860-324-9481; lostacresvineyard.com

First planted in 2010, Lost Acres has 3,000 vines of six varietals planted, including Chardonnay, Riesling, and Cayuga White. There is a nice lawn to picnic on, a deck off the tasting room, an art gallery inside, a fireplace for cold-season tastings, lots of live music events, and a modestly priced tasting with free wine glass. What's more, they have yoga at the vineyard, a beautiful pollinator garden, and a pen with pigs and horses. They even feature a wine CSA. Open Friday and Saturday 10 a.m. to 6 p.m. and Sunday 10 a.m. to 5 p.m. As a special treat, you and your sweetheart can even take a horseback ride around the vineyard, with a private picnic lunch and bottle of wine for $150. But that is so popular that they sell out months in advance, so be sure to make reservations if you plan to surprise someone for an anniversary.

Miranda Vineyard

42 Ives Rd., Goshen; 860-491-9906; mirandavineyard.com

Miranda Vineyards, another fine winery nestled in the Litchfield Hills, features wines made in the old-world traditions that owner and winemaker Manny Miranda learned from his parents in Portugal. The tasting room is located in a refurbished wood building that overlooks the vines; the rooster weathervane atop the building lends itself to the logo. Award-winning selections include single varietals—Chardonnay, Seyval Blanc—as well as Woodridge White, a blend of these two. Red offerings include Merlot and Woodridge Red, which has a pleasant earthiness along with notes of berry. Open Saturday and Sunday 12 to 5 p.m., they also feature events like live music, dinners, and fund-raisers frequently throughout the summer months and offer a chance for young and old to enjoy the scenic winery. Go to their pig roast in the fall and perhaps win a bottle of wine to take home. As they say in Portuguese, "Saude!"

Sunset Meadow Vineyards

599 Old Middle St. (Rte. 63), Goshen; 860-201-4654; sunsetmeadowvineyards.com

Sunset Meadow Vineyards has a perfect location for growing grapes—about 1,300 feet above sea level—and 21 acres of vines are planted on a western-facing slope, which you'll see as you drive up to the tasting room. The winery specializes in estate-bottled and Western Connecticut Highlands appellation selections featuring 14 grape varieties, including Cayuga White, Seyval Blanc, Riesling, and St. Croix. The Riesling has notes of lemon and a touch of sweetness. St. Croix is a beautiful red with notes of mocha and smoky spice. Twisted Red, a blend of four grapes, is medium-bodied and rich and is another among Sunset Meadow's award winners. Open year-round Monday, Thursday, and Friday 12 to 5 p.m., Saturday 11 a.m. to 6 p.m., and Sunday 11 a.m. to 5 p.m., the tasting room is a converted barn with rustic rafters and a long, wood, tasting bar. Watch swallows nest under the porch roof, or sit on the patio and look out over the vines.

Walker Road Vineyards

11 Walker Rd., Woodbury; 203-263-0768; walkerroadvineyards.com

This family-owned farm vineyard and winery specializes in blending grape varietals in classic European tradition, in an attempt to create continuity and complexity of taste, bringing out what our specific growing area has to offer. Winemaker Jim Frey looks back to places like Bordeaux, Chateau Neuf, and Chianti to see how their blending methods could succeed right here in Connecticut. Find out the results for yourself by visiting the tasting room in a renovated 150-year-old barn each Saturday and Sunday, May through December, 12 to 5 p.m.

COURTESY OF TRENA LEHN

NORTHEAST CONNECTICUT

Encompassing Tolland and Windham Counties, the Quiet Corner of Connecticut is the most rural area of the state. It is known by night-flying airplane pilots as the only "dark" area between Washington DC and Boston, the only break in the endless megalopolis suburbs that is not filled with electric light. That makes it a great place for us to find delicious maple syrup, ice cream, turkeys, and other delights. We can drive on the long scenic roads and watch the sheep and cows—and occasional llamas and bison—watch us from behind the abundant stone fences. Festivals and markets have long seasons, and pick-your-own opportunities abound here in the Last Green Valley.

On April 21, 1881, Connecticut's General Assembly accepted funding from two former Mansfield farmers, Charles and Augustus Storrs. The gift established the Storrs Agricultural College, which would eventually become the University of Connecticut. Even though they had become successful businessmen in New York City, the Storrs brothers knew how important farming was to the state, to the country, and to human culture. In an age when too many of us are separated from the land, it would probably be wise to remember their example.

Creameries and Dairy Farms

Elm Farm and Farm to Table Market

324 Woodstock Rd., Woodstock; 860-933-9594; elmfarm.net

Part of a century-old Woodstock dairy farming tradition, today's Elm Farm continues with Matthew and Christine Peckham and their four children. With 400 milking cows, beef cows, pigs, chickens, and even a few goats, they provide meat and dairy to the community of East Woodstock. Elm Farm serves its own milk and meats (beef, bacon, pork, salami, sausage) at its Farm to Table Market in Woodstock, open daily 10 a.m. to 6:30 p.m. If you see a big line, it is probably for the ice-cream window. They also carry bread and pastries, Cabot cheese, honey, maple syrup, and jam. Their apple-pie-spice bacon is a special treat.

COURTESY OF TRENA LEHMAN

Fish Family Farm

20 Dimock Lane, Bolton; 860-646-9745; fishfamilyfarm.com

This 211-acre dairy farm in Bolton uses its own bottling plant for its creamy milk, available in the store at the farm. Locally made Munson

COURTESY OF DAVID K. LEFF

Chocolate actually uses this milk for its wonderful chocolate. The ice cream is the real stand out, sold year-round by the quart in their freezer: vanilla, maple walnut, coffee, peanut butter, cherry vanilla, chocolate chip, and more. During the summertime you can enjoy their ice cream on a cone from 8 a.m. to 8 p.m. every day but Sunday. This is some of the absolute freshest ice cream you'll find, made from the milk of Jersey cows, that you can watch being pumped out every day at 3:30 p.m. Try it with some Winding Brook Farm maple syrup, made by Wayne Palmer of Hebron, available in the store. And if you've never had maple syrup on ice cream . . . well, we highly recommend it.

UConn Dairy Bar

3636 Horsebarn Hill Rd. Ext., Storrs; 860-486-2634; dairybar.uconn.edu

In business since 1954, the UConn Dairy Bar makes ice cream from the Holstein and Jersey cows at the university's farms. They feature 24 flavors at any one time, all mixed by the agricultural students. The ice cream includes 14 percent butterfat, probably contributing to its incredible popularity. The most popular flavor (perhaps due to sentimentality rather than taste) is Jonathan Supreme, named for the school's mascot, Jonathan the Husky Dog. It contains vanilla ice cream swirled with peanut butter and chocolate-covered peanuts. Other flavors include Husky Tracks, Cake Batter, and Chocolate Chip Cookie Dough. The store itself is without frills, like a classroom with stools and napkins, but it brings in 200,000 customers a year. Unlike some ice cream shops, this one is open year-round.

We-Lik-It Ice Cream

728 Hampton Rd., Pomfret; 860-974-1095; welikit.com

This 120-acre farm's herd of Holsteins is milked twice daily, and they make the ice cream fresh. It's a popular place: it's not uncommon for them to sell out, literally right after they made the batch. They often have interesting flavors; try the Guernsey Cookie if they have it, a coffee ice cream with Oreos. Road Kill is another favorite, vanilla with cherry swirl, white chocolate chips, and walnuts. The ice cream is rich and creamy, and the portions are huge. Some say to get the ice cream in a dish because the cones can't handle the amount. But the waffle cones are made fresh while you wait. Let me repeat: They make the waffle cones fresh on the spot. Open April through October 12 to 7 p.m., the farm also features hay and sleigh rides, and sells their own maple syrup and beef.

Farms and Farm Stands

Assawaga Farm

626 Providence Pike, Putnam; assawagafarm.com

Assawaga Farm is a small, diversified vegetable farm specializing in Japanese varieties of vegetables and herbs. Meaning "halfway place," the word "assawaga" is the original Nipmuc name for the river now on maps as the Five Mile. With land between the river and Mary Brown Brook, Yoko Takemura and Alex Carpenter thought this was a perfect name for their farm. Once a hayfield, this bustling little farm turns out high-quality, certified organic vegetables: cabbage and carrots, edamame and okra, tomatoes and turnips. Their Japanese offerings include shiso, komatsuna, mizuna, and Japanese green peppers. Their Sunday farm stand runs 9 a.m. to 1 p.m. from mid-May to late November.

Baldwin Brook Farm

176 Depot Rd., Canterbury; 401-419-1214; facebook.com/baldwinbrookfarm

This gorgeous farm with stone walls and green meadows hosts events like weddings, but for most of us it is just a great place to buy cuts of free-range, grass-fed heritage pork and 100 percent grass-fed Angus beef. They also have a stock of pasture-raised broiler chickens, healthy raw milk, their own branded cheeses, and a beautiful yellow spring butter. You can find Baldwin Brook in local markets and restaurants, but why not go straight to the source, since their farm store is open Monday through Saturday 10 a.m. to 6 p.m.

Bassetts Bridge Farm

552 Bassetts Bridge Rd., Mansfield; 860-455-0545; gardensatbassettsbridgefarm.com

Stroll through the formal gardens of Bassetts Bridge Farm and smell the fresh herbs. The farm is open to browse Friday through Sunday, but the self-service cart is open every day from May through October. Along with annuals, perennials, and hanging baskets of their beautiful flowers, they feature fresh fruits and vegetables during their seasons—asparagus and rhubarb in May, tomatoes and corn in July, and mums and pumpkins in September. If you gather 10 friends, in the autumn you can make an appointment for a 45-minute hayride complete with cider and donuts.

BOTL Farm

859 Westford Rd., Ashford; 908-268-319; botlfarm.com

First-generation farmers Danielle Larese and Nick Weinstock do meat right, with pasture-raised, soy-free, corn-free, non-GMO livestock, including heritage breed chickens, pigs, sheep, goats, and bees. The farm uses transitional silvopasture, with intensively managed rotational grazing, moving between 43 different paddocks throughout the year. The heritage breed animals they raise thrive in a foraging environment, and in the crazy New England climate. They offer meat and egg CSA options, and head-to-tail cuts of animals for sale. You can pick up eggs and seasonal items from 9 a.m. to 5 p.m at the honor box at the end of their drive by the farm sign. For meat, call Nick at the number above to order, or find them at several farmers markets in the area.

Burke Ridge Farms

95 Wapping Wood Rd, Ellington; 860-896-0888; facebook.com/Burke-Ridge-Farms-154836041241731

Some people in the area know Burke Ridge Farms because they sell 38 flavors of Giffords hard serve ice cream from a window at their country store. Others know it for their petting zoo. But what Burke Ridge Farms should really be known for is their Angus Beef, which they raise and show, without any hormones or antibiotics. They also sell their own pork and lamb, along with occasional goat and poultry. They are open daily from 1 to 9 p.m. In the autumn they often have pick-your-own pumpkins with hayrides and a corn maze, so check their page for updates.

Creamery Brook Bison Farm

19 Purvis Rd., Brooklyn; 860-779-0837; creamerybrookbison.net

Since 1990, a herd of bison has roamed this 100-acre farm, making it an attraction for young and old alike. Kids and adults in Connecticut are simply not used to seeing these magnificent animals, and you will probably spot them in the early evening any night of the year. The retail store on this farm is by appointment only, so call ahead to pick up this lean, antibiotic- and hormone-free protein. In past years, in July through September, you could take a wagon ride out into the woods to see them and their new calves, as well as the bewildered second-fiddle cattle that roam the property. You can even pet one of these shaggy giants. So, call ahead to arrange one of the most unique experiences in Connecticut.

Dove Hill Farm

110 Plainfield Rd., Plainfield; 203-804-1362; dovehillfarmct.com

This 7-acre family-run and -operated business prides themselves on being eco-conscious and sustainable, as well as making sure everyone has fresh, local food available. That means offering reasonable prices to customers, something that is often tough to manage for small farmers. Dove Hill grows organic vegetables like peppers, basil, and parsley, as well as fruit and flowers. Their farm stand is open 11 a.m. to 5 p.m. Tuesday through Friday, and along with the fresh produce, they sell pickles, maple syrup, and honey. Even more exciting is their selection of eggs, with chicken, duck, turkey, goose, and even quail eggs! Those who are not as food-obsessed as we are should check out their handcrafted wood products, milled right on the farm.

Ekonk Hill Turkey Farm Store

227 Ekonk Hill Rd., Moosup; 860-564-0248; ekonkhillturkeyfarm.com

The Brown Cow Cafe at Ekonk Hill has become famous for its ice cream, so they could just as easily be listed as a creamery. However, there are many reasons people come to this turkey farm, from which you can advance-order your Thanksgiving bird. The Milkhouse Bakery on the farm is also open year-round, with muffins, pies, cookies, donuts, and breads, along with the Goose and Gobbler shop, with eggs, poultry, pot pies, honey, maple syrup, and more. Of course, there are also hundreds of free-range birds to see roaming the fields. The corn maze is open in September and October, and you can also see the barnyard animals, explore the pumpkin patch, and take a hayride. At the Brown Cow, try the maple milk shake, which makes total sense and will become your new favorite. Open daily 11 a.m. to 7 p.m. during most of the year.

Fairholm Farm

72 Chandler School Rd., Woodstock; 860-923-2624; fdfairholmfarm.square.site

This century-old family farm raises 400 Holstein cows for milk, but their farm store sells their own pork and beef, along with a bevy of local products from farms like Ekonk Hill. They have a CSA of locally grown meat and offer hay wagon farm hours to see the cows get milked by robots and enjoy an ice-cream cone. The store is open Tuesday 12 to 6 p.m. and Saturday 10 a.m. to 4 p.m. Don't forget to pick up one (or more) of their signature meat rubs—we like the "low and slow" and "IPA beer" rubs in particular.

Fort Hill Farms & Gardens

260 Quaddick Rd., Thompson; 860-923-3439; forthillfarms.com

Although Fort Hill Farms contributes milk from its 200 cows to Cabot Cheese and the Farmer's Cow, it is not just a dairy farm. Active over 70 years and three generations, this agricultural destination is a place to go to appreciate the beauty and tranquility of nature. With magnificent gardens and great views, Fort Hill is open for nature walks, nature bathing, and joy-ercise. What is that, you ask? Owner Kristin Orr takes women 18 to 108 on a personal fitness workout in the gardens, fields, and forests. For those of us who like a less strenuous workout, the 7-acre corn maze opens in August for kids of all ages. Or you can sit in the garden and enjoy Fort Hill's ice cream from their herd and let the smell of lavender calm and lift your spirit.

Ghost Fawn Homestead

142 Tolland Turnpike, Willington; 662-420-0253; ghostfawnhomestead.com

This small, "beyond organic" farm in the Quiet Corner uses its resources to provide us with fresh produce while maintaining a sustainable operation. They have a small CSA and an adorable farm stand on the Tolland Turnpike, open on the weekends during the season. You'll know they have a good haul when you see the "Open" flag up. The greens here are top-notch, and you'll find quality herbs, as well.

Kollas Orchard

41 New Rd., Tolland; 860-871-0120; kollasorchardct.com

Kollas Orchard began as an experiment in 1975 and draws on David Kollas's extensive experience in the plant science department at UConn. As a fruit specialist, he uses scientific methods to determine the best growing practices, with one goal in mind: to grow great-tasting fruit. He and his wife, Janet, open the orchard every autumn, late September through late November, Wednesday through Friday 1 to 5:30 p.m. and Saturday and Sunday 10 a.m. to 5:30 p.m. They are often open in December, Friday and Saturday 10 a.m. to 4 p.m. and Sunday 12 to 4 p.m., until the fruit is gone. Their farm store is "bag your own," from the fruit they have already carefully picked and refrigerated from the trees. You will also find pies, honey, and other delights at the small farm store.

Morning Beckons Farm

343A Sand Dam Rd., Thompson; 860-821-0627; morningbeckons.com

This 250-acre estate just off Route 395 in Thompson is the largest alpaca farm in New England. Their huge herd produces show-winning alpacas in all colors and classes, and its state-of-the-art breeding program seeds many other alpaca farms throughout the region. For those of us not looking to start our own herd, their visitor center and store is open Saturday and Sunday 10 a.m. to 4 p.m. and other days by appointment. They also feature 50-minute, $10/person farm tours on which you can walk this large farm, see the mothers and babies, and spend time with the friendly herd feeding them with treat bags available in the visitor center. It's only $5 for children ages 4–11 and children under 4 are free.

Our Kids Farm

357 Barstow Rd., Canterbury; 860-208-2024; ourkidsfarmsoap.com

Owned and operated by Wendee and Neil Dupont Jr., Our Kids Farm specializes in Nigerian Dwarf goats, or rather the wonderful milk that comes from them. Goat milk has been used since ancient times as a skin cleanser and moisturizer, with alpha hydroxy acids, vitamin A, vitamin B, vitamin B12, riboflavin, and niacin. The Duponts make goat milk soap and other products to benefit the skin. "It's not just what we put into our bodies," says Wendee. "It's what we put on our bodies that's important." Their farm store and ice-cream parlor are open during the season on Wednesday through Friday 11 a.m. to 5 p.m. and Saturday to Sunday 10 a.m. to 5 p.m. Along with their goat milk products you will find eggs, meats, cheeses, milk, butter, ice cream, jams, and maple syrup among their many offerings.

Scantic Valley Farm

327 Ninth District Rd., Somers; 860-749-3286; scanticvalleyfarm.com

Raising registered hormone- and antibiotic-free Belted Galloway and Tamworth hogs, Scantic Valley Farm is a small-batch meat farm with steaks, patties, maple breakfast sausages, beer brats, and smoked kielbasa. You can also buy fresh eggs from their free-range hens, summer flowers, and pick-your-own berries. Often open 8 a.m. to 2 p.m. for meat pick-up and pick-your-own, call ahead to make sure of the correct times, since Scantic Farm waits until their products are just right for us to harvest.

Shundahai Farm

253 Maple Rd., Mansfield; 860-429-0695; shundahaifarm.com

Since 2009 Raluca Mocanu and Ed Wazer have run this small, 5-acre family farm only a mile away from the University of Connecticut. They are dedicated to sustainable practices, never spraying pesticides, herbicides, or any chemicals on the farm. At their farm stand, which is open Friday 3 p.m. to 6 p.m., you can find their garden vegetables, peaches, apples, and both European and Asian pears. They sell organic, non-GMO, soy-free grain-fed chickens and 100 percent grass-fed beef at certain times of year, but they sell out fast, so keep your ear to the ground, or better yet, join their mailing list. Shundahai means "Peace and harmony with all creation" in Shoshone, which captures the spirit of these intrepid farmers perfectly.

COURTESY OF TRENA LEHMAN

Stearns Farm Stand

483 Browns Rd., Storrs; 860-382-2303; facebook.com/stearnsfarmstand

The Stearns family farm runs a daily farm stand that sells local native fruit and vegetables—corn, tomatoes, apples, peaches, eggplants, squash, and more—along with meat, bread, and dairy products like yogurt, milk, and ice cream. You can browse a selection of jams, pickles, honey, and maple syrup, or just run in to pick up your favorite pie. Unique offerings like Bloody Mary mix and cream cheese spreads make this a one-stop shopping experience, and the daily 10 a.m. to 6 p.m. stand make Stearns as convenient as a grocery store. But with better food.

Tardif Farm

89 Flanders Rd., Coventry; 860-498-0599; tardiffarmandfeed.com

This is a poultry farm that is more than a poultry farm. The store at Tardif farm sells corn, tomatoes, cucumbers, squash, potatoes, peppers, zucchini, eggplant, bread, cookies, pies (fresh every day!), milk, wool, yarn, crafts, honey, jams, jellies, handmade soap, maple syrup, spices, sauces, and handmade pottery. As their website name implies, they also sell animal feed, mulch, and other farm needs, for those of you starting your own agricultural business. The rest of us can visit on special farm tour days and events like "Cornstalks and Crafts," and the store is open Wednesday through Friday 10 a.m. to 6 p.m., Saturday 10 a.m. to 5 p.m., and Sunday 10 a.m. to 6 p.m.

Taylor Spring Farm

558 Buckley Highway (Rte. 190), Union; 860-949-2575; taylorspringfarm.com

This family farm humanely and sustainably raises beef cattle, chickens, turkeys, and pigs all on pasture. The small herd of beef cattle enjoys year-round access to pasture, shelter, fresh water, and local hay. The pigs roam through a mix of forest and meadow, rooting for shoots and acorns, as well as being fed a locally milled whole grain mix. Their Cornish Cross chickens are moved to fresh pasture daily and their turkeys explore the fields daily. These are all available in their farm store except the 15–25-pound turkeys, which sell out quickly every autumn, so place your order for Thanksgiving early. The farm also keeps bees, which produce a limited amount of honey that is also available every year. The store is open Saturday and Sunday 9 a.m. to 3 p.m.

The Country Butcher at Spring Meadow Farm

1032 Tolland Stage Rd., Tolland; 860-875-5352; countrybutcherct.com

This country butcher shop run by the Boyer family at Spring Meadow Farm is a great example of how doing things the old-fashioned way can be the most "gourmet" way of all. Their meats, artisan grains, and specialty products at the farm store are sourced from their own 80-acre farm and other local and regional providers and craftspeople. But it is their own meticulous craft of butchering meat that makes this such a special place. Prime certified beef, natural, grain-fed pork, and organic, antibiotic free poultry is always good, but when prepared by a skilled craft butcher, it is taken to another level. The store is open Friday 10 a.m. to 6 p.m. and Saturdays 9 a.m. to 4 p.m. Try their gourmet sauces, artisan baking mixes, or one of their 30 types (!) of hand-made sausages.

Willow Valley Farm

39 Moose Meadow Rd., Willington; 860-508-7706; wvfcsa.wordpress.com

"Good for the Earth. Good for you." That is the motto of the folks at Willow Valley, a third-generation family farm with 64 acres along the Fenton River. Once a commercial chicken farm, they diversified into cows and pigs, blueberries, and summer camps. As "stewards of the land," they practice sustainable growing methods and have taken the Connecticut Northeast Organic Farming Association pledge. They concentrate on getting the right balance of plants, nutrients, and microorganisms to keep healthy soil, while using natural products like garlic spray for pests and pathogens. Furthermore, they allow wild plants to grow on the edges of the fields, something that to a casual viewer might look "messy," but in fact is a sign of a truly vigorous ecosystem. Their farm stand is open Saturday and Sunday 9 a.m. to 3 p.m. Pick up their healthy produce self-serve with cash or check, and eat in good health.

Farmers Markets

Ashford Farmers Market

25 Tremko Ln., Ashford; facebook.com/ashfordfarmersmarketct

Set in Pompey Hollow Park, the Ashford Farmers Market runs rain or shine from April 25 through October, Sunday 10 a.m. to 1 p.m. You'll find all the usual farmers market bounty: fresh produce, eggs, baked goods, mushrooms, maple syrup, and local honey. But along with live music, herbal products, and local crafts, they often feature massage therapy. Some might argue that going shopping for fresh produce on a cool spring day is therapy enough. But not us! Farmers markets are a great way to sample the products and services of farms that are not usually open to the public, and to discover new favorites along the way.

Coventry Regional Farmers Market

2299 South St., Coventry; 860-742-1419; coventryfarmersmarket.org

Coventry boasts the largest farmers market in the state from June to October every Sunday 10 a.m. to 1 p.m. The setting at the Nathan Hale Homestead is a gorgeous and historic backdrop for what is really more a small fair than a farmers market, with horse and wagon rides, demonstrations, sheep shearing—you never know! Fiddlers provide music every weekend, and dozens of vendors show up selling everything from traditional jams and preserves to smoked bacon to chocolate fudge. You'll also find local artists and artisans selling handmade wares, from soap to candles to beadwork. And if you're in the mood for a market from December to March, Coventry runs one of those too, at the same time of week and day but at the Community Center, 124 Lake Street, Coventry, with about 30 farmers and vendors taking part. After all, we don't stop eating in winter.

Ellington Farmers Market

Arbor Park, Main St., Ellington; ellingtonfarmersmarket.com

Like its name implies, Arbor Park has numerous shady trees spread out over green lawns, and this makes browsing the vendors at the Ellington Farmers Market a delightful experience, even on a hot

COURTESY OF TRENA LEHMAN

summer day. You can enjoy this dog-friendly market and its live music every Saturday from July through October, 9 a.m. to 12 p.m. Sample Caribbean-style treats from the Amazing Ackee and crafts from Muddy Brook Potters, along with produce, dairy, meat, and more from the farms of Tolland County. Check their schedule for events like the Peach Festival or the Salsa Festival—every weekend is something new. This market supports low-income families through matching SNAP and Farmers Market Nutrition Program vouchers, so not only are you supporting local farmers and craftspeople, but you are also helping to sustain the neediest among us every time you visit Ellington on a Saturday morning.

Northeast Connecticut Farmers Market

564 Providence Rd., Brooklyn; 860-564-1117; nectfarmersmarket.org

Though northeast Connecticut is the most rural area of the state, you might wonder why there are fewer farm stores. One reason might be because there are so many high-quality farmers markets in the area. A great example is the Northeast CT Farmers Market Association, which has operated farmers markets in the Quiet Corner of the state since 1980. These days, they have not one but four centralized locations for people like us to find fresh produce and farm products, in Killingly, Putnam, Plainfield and Brooklyn. They run the markets from May to October and even have a CSA that consumers can pick up weekly at the Putnam location from 4:30 to 6 p.m. The Brooklyn Commons Shopping Center location listed above meets Wednesday 4 to 6 p.m.; another meets at the Killingly Public Library, 25 Westcott Road, Saturday 9 a.m. to 12 p.m.; a third at the Plainfield Early Childhood Center, 651 Norwich Road, Tuesday 4 to 6 p.m.; and the last at the Putnam Riverview Marketplace, 18 Kennedy Drive, Monday 3:30 to 6 p.m.

Putnam Saturday Farmers Market

18 Kennedy Dr., Putnam; 860-963-6834; putnamfarmersmarket.org

Located on the banks of the Quinebaug River under the pavilion of the Riverview Marketplace, the Putnam Saturday Farmers Market is sponsored by the town's Office of Economic and Community Development. They know how important these events are not just for local businesses and customers, but for community morale. A town with a farmers market is a town with a sense of self. Open Saturday, early June through late October, 10 a.m. to 1 p.m., the market brings you everything from lettuce to lilies to llama wool yarn. And it's right next to the Putnam Lions Memorial Dog Park, so bring your furry friend along for a romp. Or walk the other direction along the Putnam River Trail to see the always impressive Cargill Falls.

Storrs Farmers Market

4 South Eagleton Rd., Storrs; storrsfarmersmarket.org

Founded in 1994, the Storrs Farmers Market brings locally grown foods to the students and workers at UConn, and the rest of the town of Mansfield. The market features only Connecticut farmers, like maple syrup and honey producer George Bailey, with the average distance from the market being a mere 10 miles. The market runs at the Mansfield Town Hall from May to November, Saturday 3 to 5 p.m. But our bodies appreciate fresh food all year round, and farms don't shut down in the colder months. So they also run a winter farmers market, which is open twice a month (first and third Saturday) from December through April, also 3 to 5 p.m. in the Buchanan Auditorium of the Mansfield Public Library.

Willimantic Farmers Market

28 Bridge St., Willimantic; 860-423-0533; willimanticfarmersmarket.org

The oldest running farmers market in Connecticut, Willimantic opened over 45 years ago with a mere 15 local vendors and a cooperative association to ensure its continuing quality. By 2016, the market had expanded and moved to a larger open space on Jillson Square, open Saturday 8 a.m. to 12 p.m. at the Willimantic

Whitewater Park. Open May until the end of October, you can get asparagus and spinach in May, peppers and plums in July, or beans and beets in October. There's a good reason this market has lasted so long, and that is because of a commitment to quality food by quality farms.

Farm Stays

Henrietta House Bed and Breakfast

125 Ashford Center Rd., Ashford; 860-477-0318; historichenriettahousebnbct.com

This 1700s house with traditional center chimneys and five working fireplaces sits on a 3-acre farm in Ashford. With its hand-sawn paneling and beehive oven, the house makes you feel like you've stepped back in time. However, the accommodations are renovated, with deep soaking tubs and heated floor tiles. Still, you are probably coming here for the rambling stone walls and 1800s barn. You can roam the landscape and eat fresh produce from the large garden or

COURTESY OF TRENA LEHMAN

blueberries from the field behind the barn. There are goats, chickens, and occasionally pigs, all of which provide the fantastic breakfasts you might have here at Henrietta House.

Woodstock Sustainable Farms & The Manton-Green Bed & Breakfast

211 Pulpit Rock Rd., Woodstock; 888-788-8726; wssfarms.com

It's hard to know where to start with this miraculous 200-acre "farm for the future." You might be impressed by the sustainable initiatives, from an on-site DC microgrid to an integrated biochar heating system. You might want to stop by their retail store Saturday 10 a.m. to 6 p.m. to pick up some of their pasture-raised beef, chicken, pigs, and lamb, as well as heirloom vegetables, eggs, and more. You could make a reservation for their weekly hearth dinners (cooked in a traditional 18th-century fireplace) to taste timeless New England fare like beef stew, lamb stew, hearth-baked bread, and peach cobbler. Or, you might want to experience their bed and breakfast, staying in remote locations on the farm in either the 1710 Manton-Greene House or the 1730 Kingston Herb Barn. The Manton-Greene House has undergone a museum-quality restoration, with hidden luxuries like radiant heat and a steam shower. While staying at Woodstock Sustainable Farms, you will enjoy their own eggs, bacon, and fresh beehive-oven-baked bread. You can also schedule a massage, reflexology appointment, or Reiki session. The geniuses at this farm have somehow hit the balance between old and new, and anyone who stays here will come away with an appreciation for how humans might not just survive but thrive in the future.

Festivals and Fairs

Brooklyn Fair

15 Fairgrounds Rd., Brooklyn; 860-779-0012; brooklynfair.org

Founded in 1809, the Brooklyn Fair is the oldest continuously operated agricultural fair in the United States. The Windham County Agricultural Society hosts four days of fun the weekend before Labor Day, with people from all over New England coming to this small Connecticut town to see oxen pulls, draft horse demonstrations, skillet tossing, dog shows, cattle parades, and more. The circus and midway games entertain the kids, while the bingo and barbecue bring in the older folks. Everyone enjoys the country dancing, art shows, and beekeeping exhibits. There are craft vendors to bargain with and food vendors to gobble from. The fair is free, but of course the rides, games, crafts, and food are not.

Connecticut Sheep and Wool Fiber Festival

Tolland Ag Center, Tolland;
ctsheep.org/ct-sheep-wool-fiber-festival

The original Constitution and By-Laws of the Connecticut Sheep Breeders Association said, in part, "The purpose of the Association shall be to promote and encourage the keeping of sheep upon the farms of Connecticut, to improve breeds of sheep, and to aid in securing legislation favorable to the sheep industry and agriculture." That was 1893, and the Association is still mentoring shepherds and providing support to sheep farmers through the Connecticut Blanket program. One of the best things they do is the annual Sheep and Wool Fiber Festival, held since 1909. On the last Saturday in April, at locations around the state, but often in Tolland, this family-friendly agricultural and educational experience is a favorite of shepherds and fiber enthusiasts, but also for the rest of us.

Hebron Harvest Fair

347 Gilead St., Hebron; 860-228-0542; hebronharvestfair.org

Growing steadily since 1971, the Hebron Harvest Fair is today one of the largest agricultural fairs in Connecticut. Produced by the Hebron Lions Agricultural Society, this four-day extravaganza uses its proceeds to fund grants for local and national charitable causes. It is a classic country fair, with food and farm stands, oxen and pony pulls, demolition derby and tractor pulls, music and midway. The diaper dash is a favorite for new parents, and the chick hatchery is a favorite for the kids. Hay maze, pie-eating contest, handmade quilts, and a full-on agricultural museum . . . this is an agricultural fair that has something for everyone. Children 12 and under, and active military get in free, with discounts for other groups, as well.

It takes place every September, Thursday 4 to 10:30 p.m., Friday 12 p.m. to midnight, Saturday 9 a.m. to midnight, and Sunday 9 a.m. to 7:30 p.m.

Hebron Maple Festival

Main St., Hebron; 860-428-7739; hebronmaplefest.info

This sweet, sweet festival takes place throughout the town of Hebron on the second weekend in March every year 10 a.m. to 4 p.m. both days. All the local sugarhouses give tours, but that is only the tip of the maple leaf. Enjoy a craft fair, blacksmiths, candle making, woodworking, face painting, an antique tractor parade, an ice-cream eating contest, pancake breakfasts, and more. You can eat icy maple milk, hot dogs, homemade soup, glazed donuts, kettle corn, fried dough, and maple pudding cake. There is little else going on this time of year, so why not head out to Hebron and enjoy the end of the maple sugaring season in style?

Woodstock Fair

281 Rte. 169, Woodstock; 860-928-3246; woodstockfair.com

Always on Labor Day weekend at the Woodstock Fairgrounds, this fair is open Friday through Sunday 9 a.m. to 10 p.m. and Monday 9 a.m. to 6 p.m. It has been going strong since it began during the height of the Civil War. Every year 200,000 people come to tiny Woodstock to enjoy crafts, food, musical entertainment, livestock shows, a petting zoo, stage shows, go-kart races, strolling entertainers, and much more. The lights of the midway are spectacular; this is a fair that definitely comes into its own during the evening hours.

Maple Sugarhouses and Apiaries

Bright Acres Farm Sugar House

46 Old Kings Hwy., Hampton; 860-455-9654; facebook.com/BrightAcresFarmSugarhouse

Richard and Judy Schenk sell maple syrup and honey year-round, along with chicken and duck eggs, from their small family farm on Old King's Highway. Their high-quality maple syrup goes quickly, though, so try to get it in season just after the February-March sugaring season. Bright Acres holds an annual open house in March, and that is the time to visit and watch the always-fascinating process. They will teach you the history of syrup making, from its earliest beginnings in pre-colonial America to the latest technological advancements. They'll show you the difference between sap buckets and plastic tubing, and you'll even get the chance to tap a tree. The Schenks have grown this operation from a small hobby to a full-fledged business, and they have done it by paying attention to best practices. That leads naturally to high-quality maple products for us to enjoy.

Hurst Farm Sugar House

746 East St., Andover, 860-646-6536; facebook.com/Hurst-Farms-626002090881388

Hurst Family Farm has a lot to offer: a CSA from June to October, fall harvest hayrides, and a beautiful country store full of their own jams, jellies, relishes, maple, and honey. But the 36-acre farm's star might be their post and beam sugarhouse, (handicapped accessible!) which is open to the public during February and March. Call for hours to watch the sugar-making process, but the farm store is open year-round, Monday through Friday 9 a.m. to 5:30 p.m. and Saturday and Sunday 9 a.m. to 5 p.m.

Hydeville Sugar Shack

118 Hydeville Rd., Stafford Springs; 860-916-9645; hydevillesugarshack.com

The Hartenstein family operates their wonderful sugar shack for six weeks every year, tapping, collecting, and boiling down the sap into a delightful and dark maple syrup. You can call ahead for a visit, usually available in the afternoon in season during February and March. You can also find them and their syrup, candy, cream, and sugar at local farmers markets and events. "For the Love of all things Maple," is the Hartenstein's motto, and it's a sweet one.

Norman's Sugarhouse

387 County Rd., Woodstock; 860-974-1235; rnorman4.wixsite.com/mysite

In the 1970s, Richard and Avis Norman began making maple syrup as a hobby, using a cooking pan and fire pit and producing enough for a few friends. Since then, they have expanded to a year-round operation with their product in stores and restaurants around the state and beyond. They continue to make the syrup step-by-step, or rather, hand-by-hand, boiling it down and canning it. They also make maple candy and cream, and help eager young maple tree tappers in their own journey. If you want to visit for a tour or buy their delicious product, call ahead.

River's Edge Sugar House

326 Mansfield Rd., Ashford; 860-429-1510; riversedgesugarhouse.com

This rustic, family-run sugarhouse on the Mount Hope River off Route 44 has been tapping since 1993, though they have expanded a lot since then. With a sugarhouse, evaporator, reverse osmosis machine, and thousands of tapped trees, River's Edge makes some of the best maple syrup in the state. They also keep a number of beekeeping hives and produce their own honey. With a full kitchen and storefront, this farm family allows you to watch their process in February and March, as 50 gallons of sap becomes 1 gallon of pure syrup. Their store is open year-round, but call ahead to make sure they are there to serve you.

Museums and Education

Blue Slope Country Museum

138 Blue Hill Rd., Franklin; 860-642-6413; blueslope.com

On working days, Blue Slope is a trucking company, moving farm commodities like milk, hay, grain, and sawdust. But it doubles as a small country museum and is full of historical farming and country life artifacts. A 200-year-old stone spring house, a bank barn that houses four Belgian draft horses, and a museum with hundreds of antique tools are some of the attractions on this 380-acre dairy farm. They offer educational programs, events, and activities for groups of children or adults, from horse-drawn wagon rides to family campfires to square dances. Look at their website for the schedule and call for details.

Strong Family Farm

274 West St., Vernon; 860-874-9020; strongfamilyfarm.org

Some people know the Strong Family Farm for their store, where they find local produce like organic rhubarb and preserved goods like blackberry jelly. (Open Wednesday to Friday 1 to 5 p.m., Saturday 10 a.m. to 2 p.m., and Sunday 11 a.m. to 3 p.m.) After all, the 50-acre Strong Family Farm has been in business since 1878, long enough to last for seven generations of Strongs. Over the years there have been many changes, like from dairy farm to turkey farm in the 1960s, but the biggest change came in 2010 when owner Norman Strong died. His daughter Nancy decided to make the farm an educational nonprofit to serve the town of Vernon as an agricultural education center. They host various activities throughout the year, including a 5K Chicken Run the second Sunday in April, a

COURTESY OF TRENA LEHMAN

chicken care program during the summer months, animal themes on Sunday and a harvest festival the third Saturday of October. Educational programs, classes, demonstrations, and lectures for both children and adults give everyone a farm experience, but also demonstrate the power and importance of farming and gardening for the future health of humanity.

Pick Your Own

Buell's Orchard

108 Crystal Pond Rd., Eastford; 860-974-1150; buellsorchard.com

Just off Route 198 in Eastford, this 100-acre farm has pick-your-own fruit, beginning with strawberries in June and then moving through blueberries, peaches, apples, and pumpkins. Their small farm store also sells their harvested produce, as well as cheeses, preserves, candles, maple syrup, and cider. Buell's has become famous for its caramel apples, available after Labor Day. From August to October, watch the apples enter the machine and become candied. Check the website for updated seasonal hours, usually Monday through Saturday from June through October. Check for the day they host an annual fall festival, with hayrides, apple pie, hot dogs, hamburgers, cider, and donuts. This is the most popular farm in the area, and everyone speaks fondly of their food and their fun.

DeFazio Orchard & Greenhouses

1393 North Rd., Dayville; 860-774-3799; facebook.com/DeFazioOrchardandGreenhouses

DeFazio Orchard and Greenhouses is a family-owned farm growing fresh fruits, vegetables, and flowers, with an incredible selection of hanging baskets and pots. You can come and enjoy pick-your-own fruits seasonally, including blueberries, peaches, and apples. Free wagon rides are available for the kids. Defazio is open 9:30 a.m. to 5 p.m. every day except Sunday, when it is open until 4:30 p.m.

Horse Listeners Orchard

317 Bebbington Rd., Ashford; 860-870-7301; horselistenersorchard.com

This gorgeous 153-acre orchard includes a large selection of apples, peaches, tomatoes, vegetables, and blueberries all ripe for the picking. Their farm store is open Monday through Thursday 9 a.m. to 4 p.m., and Friday through Sunday 10 a.m. to 6 p.m. They stock both their own products and ones from other local farms, including cheese, butter, honey, apple cider, canned jams and jellies, pickles, pies, applesauce, and eggs. They feature train rides on weekends, as well as other tours and monthly events throughout the season.

Irish Bend Orchard

90 Pioneer Heights Rd., Somers; 860-698-6429; acebook/irishbendorchard

The highlight of this fourth-generation farm in Somers is an orchard of apples, Asian pears, peaches, and pumpkins, all available for pick-your-own. They also grow heirloom vegetables and other items like squash and gourds. The u-pick orchard and farm stand are open August through October, Friday to Sunday 10 a.m. to 6 p.m. In the autumn kids of all ages love taking a hayride to the pumpkin patch.

COURTESY OF TRENA LEHMAN

Lapsley Orchard

403 Orchard Hill Rd., Pomfret Center; 860-928-9186; lapsleyorchard.com

General Israel Putnam owned this area in 1750, raising stock and growing fruit between the wars he served in with his friend George Washington. Today, this 200-acre farm is owned by John Wolchesky, a first-generation farmer who took over Lapsley in 1984. He diversified the offerings from pear trees, adding peaches and berries and a 60-acre "garden" of vegetables. A retail stand offers their selection of fruits and vegetables along with other local delights.

In fact, Wolchesky pioneered this direct farm-to-consumer market plan back in the 1990s. You can also join their eight-week CSA and, of course, enjoy the pick-your-own blueberries, apples, pumpkins, and flowers. They have horse-drawn wagon rides and a fall festival on Columbus Day weekend, with live music, cider slushies, and food cooked by the Pomfret Lions Club. The farm stand opens in mid-July daily 10 a.m. to 6 p.m. and closes on Christmas eve. Don't miss their amazing cider donuts, made with their own fresh pressed cider and appley beyond compare. "Time slows down when you're on the farm," says Wolchesky. It is so slow at Lapsley Orchard you might not want to leave.

Raspberry Knoll Farm

163 North Windham Rd., North Windham; 860-786-7486; raspberryknoll.com

Raspberry Knoll Farm is a 65-acre family-run farm, owned by Mary and Pete Concklin, specializing in pick-your-own berries, vegetables, herbs, and flowers. They have 10 varieties of raspberries alone, along with blackberries, strawberries, and blueberries. Herbs like basil and cilantro, flowers like asters and zinnias, and vegetables like melons and onions all make appearances in the fields and in the farm store. The cash-only farm is open Wednesday 9 a.m. to 5 p.m., Thursday 9 a.m. to 7 p.m., Friday 9 a.m. to 5 p.m., Saturday 8:30 a.m. to 5 p.m., and Sunday 9 a.m. to 5 p.m. They also feature garden classes, so check their website for these special events.

Woodstock Orchards

494 Rte. 169, Woodstock; 860-928-2225; woodstockorchardsllc.com

Raised on the famous Bishops Orchard of Guilford, Harold Bishop bought Woodstock Orchards in 1958. Now owned and operated by the next generation, Douglas and Donna Young, the farm has wisely diversified, added greenhouses, and updated its harvesting methods. Today the third generation has started a bakery, which makes pies, muffins, apple crisps, and apple cider donuts (made by a robot that the kids love to watch). They feature pick-your-own apples and blueberries in season (check their website), and farm-fresh vegetables like cucumbers and garlic in the summer. They grow peaches and plums, too, but offer those only in the store. The store is open every day from 9 a.m. to 5 p.m. and accepts cash, check, debit, and credit. The 20 different apple varieties are sold by the bag, although the bags come in different sizes! You can pick a peck if you like.

Wright's Orchard & Dried Flower Farm

271 South River Rd., Tolland; 860-872-1665; wrightsorchard.com

Wright's Orchard started in 1981 with a small planting of 250 semi-dwarf trees. Since then, the orchard and the store have both expanded considerably. Their pick-your-own includes a wide variety of apples, blueberries, and pumpkins. You can buy fresh peaches, raspberries, and vegetables at the farm store, and pick out a bouquet from the dried flower barn! They also have hardy mums, pumpkins, gourds, winter squash, tomatoes, apple cider, plums, and more. They prepare fruit pies and make apple cider donuts. If you need a jack-o'-lantern for your doorstep, this is the place to come. The farm is open from July through December—with pick-your-own open from approximately August 20 until October 31—Monday to Saturday 12:30 to 5:30 p.m. and Sunday 1 to 5:30 p.m.

COURTESY OF TRENA LEHMAN

COURTESY OF TRENA LEHMAN

Trail Rides

Shallowbrook Equestrian Center

247 Hall Hill Rd., Somers; 860-749-0749; shallowbrook.com

This 50-acre farm is one of the largest family-owned horse complexes in America, and one of its top riding schools. Shallowbrook has a hunt course, indoor and outdoor rings, an outdoor polo course, and the largest indoor polo arena in the country. They offer riding lessons for all levels and ages during all seasons of the year, and their special birthday party events are hugely popular. Check their events, because rodeo, carriage shows, and more happen here, and Shallowbrook becomes not just a place for you to be active but also a venue for spectator sports like polo and championship horse shows.

Valley View Riding Stables

91 Lake Rd., Dayville; 860-617-4425; valleyviewridingstables.com

This full-service equestrian center in Dayville was once a dairy farm but is now occupied by owner Amy Lyons and more than 30 horses. She has installed a large indoor riding arena and a huge outdoor arena that includes a 12-fence hunter course, a 30-obstacle cross-country course, and a 60-foot round pen. A variety of trails snake through the 250 acres of woods and pastureland, and you can go on guided trail rides on these after a 30-minute lesson. For younger children there are pony rides and birthday parties. Just be sure to wear jeans and tough shoes, but otherwise they provide all the equipment. The prices are very reasonable, and you'll be amazed at how an hour on a horse will change your entire outlook on life.

Tree Farms

Cedar Ledge Tree Farm

260 Coventry Rd., Mansfield; 860-423-5690; cedarledgetreefarm.com

Nestled in the scenic hills of Mansfield, Connecticut, Cedar Ledge has been operated by the Cone family as a tree farm since 1983. Christmas trees were only supposed to be a hobby for Ken Cone, but by 1992 when the first harvest took place, Cedar Ledge began to grow this seasonal business. A Christmas shop sells wreaths and garlands, and now they have over 10,000 trees, with more being planted every spring. The farm is still managed for other recreation, too, with tractor rides, mulch mountain slides, pedal karts, and pumpkin picking 10 a.m. to 5 p.m. on weekends from Columbus Day to Thanksgiving, with tree cutting following, 9 a.m. to 5 p.m. on the weekends and 12 p.m. to 5 p.m. on weekdays between Thanksgiving and Christmas. They sell stone, too, if you are in the market for it!

COURTESY OF TRENA LEHMAN

Harmony Farm Trees

294 Bedlam Rd., Chaplin; 860-428-7398; harmonyfarmtrees.com/home

A good reminder that a Christmas tree is more than just any old product is that Harmony Farm first planted them in 1986 and didn't start harvesting them until 1995. Steven Laume and his team offer 6 acres of White Spruce, Blue Spruce, White Pine, Meyer Spruce, and Fraser Fir for you to cut, Friday to Sunday the day after Thanksgiving to Christmas Eve. Kids are given a free candy cane and coloring book, and the whole family can sit by the woodstove in the barn and warm up after the adventure.

Hartikka Tree Farm

262 Wylie School Rd., Voluntown; 860-376-2351; treeman2.com

This family business has been growing Christmas trees since 1955, with ten varieties of quality trees. Most years they offer horse-drawn cart rides, wood-fired pizza, hot chocolate, and donuts to satisfy you while you visit. You can buy fresh hand-crafted wreaths and garland, and other accessories for your freshly cut tree, as well. They open after Thanksgiving and take cash or checks only, weekdays 10 a.m. to 5 p.m. and weekends 9 a.m. to 5 p.m. If you are in the market for a truly enormous tree, a 10-to-15-footer, this is the place to come. Just call ahead and reserve one of these special giants.

Hickory Ridge Tree Farm

108 South River Rd., Coventry; 860-918-3416; hickoryridgetreefarm.com

Hickory Ridge is a fourth-generation farm named for the many hickory trees growing on the hills here. Since 1960 they have specialized in Christmas trees, with 12 varieties, including Fraser Fir, Canaan Fir, and Blue Spruce. Their staff will help you pick out the best for your particular needs. For example, do you need strong branches or better needle retention? They are usually open Thanksgiving to Christmas on Wednesday and Thursday 12 to 5 p.m. and Friday, Saturday, and Sunday 9 a.m. to 5 p.m. The 50 acres of this farm contribute to one of the largest unbroken forests in Southern New England. Come pick out your tree from that forest and know that a new one will be planted in its stead.

Pulpit Rock Nursery

307 Pulpit Rock Rd., Woodstock; 860-377-2374; pulpitrocknursery.com

This 25-acre Christmas tree farm just up the road from Taylor Brooke Winery is a family affair, with the Kaeser family managing the hills of green and blue trees. You can bring your own bow saw or chainsaw, along with cash or checks (no credit cards) and enjoy a day out in the chilly air with your family. Open starting the Friday after Thanksgiving, and Saturday and Sunday between Thanksgiving and Christmas 9 a.m. to 4 p.m.

Wineries

Cassidy Hill Vineyard

454 Cassidy Hill Rd., Coventry; 860-498-1126; cassidyhillvineyard.com

The lovely town of Coventry in the Last Green Valley National Heritage Corridor provides the backdrop for Cassidy Hill Vineyard. The picturesque landscape spreads out over 150 acres, ample room to stroll and enjoy the vines. The winery's logo was inspired by a lone maple called the Thinking Tree, located just a short walk from the log cabin tasting room. Balanced between whites and reds, the tasting menu includes a white wine infused with strawberries called Summer Breeze and a blend of Merlot and estate-grown St. Croix called Coventry Spice. The tasting room is open 11 a.m. to 8 p.m. on Friday and 11 a.m. to 5 p.m. Saturday and Sunday.

Heartstone Farm & Winery

468 Rte. 87, Columbia; 860-337-0162; heartstonewinery.com

This beautiful farm once grew berries and raised cattle, but added wine grapes in 2008, opening the winery to the public in 2017. They plant and bottle Cayuga White and Corot Noir, Frontenac and Marquette, cold-hardy varietals that grow well on their 600-foot hilltop property. You can taste and sip and enjoy food trucks or a picnic lunch. Open April to November, Thursday 12 to 7 p.m., Friday 12 to 9 p.m., Saturday 12 to 7 p.m., and Sunday 12 to 6 p.m. "Come enjoy with us the fruits of our farm," say owners Walt and Nancy. It's a good reminder: sometimes we forget that wine grapes are fruit.

Sharpe Hill Vineyard

108 Wade Rd., Pomfret; 860-974-3549; sharpehill.com

Hidden in the hills of Pomfret, past forests and meadows, you'll see a western rail fence and the huge red barn of one of Connecticut's largest and most award-winning wineries. Owners Steven and Catherine Vollweiler and the winemaker, Howard Bursen, have created an institution here in this barn, which contains the winery itself, the barrel room, the tasting room, and their unique restaurant, the Fireside Tavern. They produce the best-selling wine in New England, Ballet of Angels—a semisweet summer pleaser, cool and refreshing. But their other wines are even more worthy of attention, such as the *Wine Spectator*–rated Chardonnay and the St. Croix, which tastes like spiced roses. Sit on the stone terrace behind the tasting room and enjoy a glass, but you should certainly take a walk up the

hill through the rows of grapevines. At the top you'll have a view of Windham County and beyond into Massachusetts and Rhode Island. Open for tastings and sales Friday through Sunday 11 a.m. to 5 p.m.

Taylor Brooke Winery

848 Rte. 171, Woodstock; 860-974-1263; taylorbrookewinery.com

Owners Dick and Linda Auger want visitors to feel at home when they arrive at Taylor Brooke Winery. The winery and tasting room are found in the beautiful landscape of the Quiet Corner, and a pleasant drive through Woodstock will take you there. A wide variety is offered, including intriguing blends featuring fruit infusions. Try Riesling on its own, then sample the wines that blend Riesling with concentrated essences—Green Apple Riesling, Summer Peach, Autumn Raspberry, and Cranberry Riesling are all surprising and fun. Don't miss Chocolate Essence, a Merlot port infused with chocolate: "dessert in a glass." Taylor Brooke pioneered the use of the Traminette grape, and the wine made from these hybrid grapes is delightfully floral and smooth with just the right touch of sweetness. Their version of Cabernet Franc is aged in Hungarian oak and has hints of mocha on the finish. They also have a brewery (taylor brookebrewery.com), so if you prefer an India pale ale, you can see what's on tap, as well. Open Friday 2 to 6:30 p.m., Saturday 12 to 6:30 p.m., and Sunday 12 to 5 p.m. Bring lawn chairs and set up a blanket between the rows of vines.

The Vineyard at Hillyland

75 Murphy Hill Rd., Scotland; 860-786-7770; thevineyardathillyland.com

This former 300-acre dairy farm owned by the Stearns family began growing grapes in 2007 and remarkably produces all estate-grown wines. They have live music on weekends, which you can enjoy in their tasting room, converted from a century-old saltbox-style garage. Their wines include a dry, floral Traminette and an earthy, smooth St. Croix. Our favorite might be the Ridgedale, made with Marechal Foch grapes, a red wine with sublime chocolate notes. Tastings and sales are open Friday 3 to 8 p.m. and Saturday and Sunday 12 to 5 p.m., with longer hours during the summer season.

SOUTHEAST CONNECTICUT

In 1760, the Reverend Jared Eliot of Killingworth published the first book of agricultural guidance in the American colonies, *Essays Upon Field-Husbandry in New-England, as it is or may be Ordered*. "A discovery of the nature and property of things and applying them to useful purposes, is *true philosophy*," he wrote. And no one knows that true philosophy better than farmers.

Today, the farms in Eliot's corner of the state are part of the "playground" of Connecticut, with tourist attractions that draw thousands of visitors every year. And yet, agricultural tourism opportunities are a big part of that draw. It is a varied land—Middlesex County hugs the Connecticut River, with rocky hollows and hills flanking it, while New London County's rural inland areas seem radically different from the beach villages and harbors of the shore. You can taste local Noank oysters or Jeremy River cheddar cheese, washed down with the clean, crisp chardonnays of the Southeastern New England AVA. This area of Connecticut features some of the prettiest towns in America, like Essex and Stonington, but getting out of those towns into the farmlands beyond is part of what makes us return again and again.

Creameries and Dairy Farms

Brush Hill Dairy

87 Brush Hill Rd., Bozrah; 860-383-9255; brushhilldairyllc.wixsite.com/brushhilldairyllc

The delicious raw cow's milk from Brush Hill Dairy comes from a herd of Holsteins, Jerseys, and Dutch Belts that graze on land originally purchased by the Brush family in 1887. Back then, it was a dairy farm, too, selling butter to local markets, but eventually the original herd was sold. For a number of years, the family continued to "hobby farm," renting out their cropland, until 1994, when Sarah Brush and her husband, Texas Moon, decided to pursue their dairy farm dreams and become the fourth generation to work the land purchased by Sarah's family over 100 years earlier. Their products go beyond raw milk to pork, vegetables, veal, and greenhouse plants, available in a 30-member CSA or at their farm store. During the summer, find them at the Bozrah farmers market on Friday from 4 to 7 p.m., and at the Waterford farmers market on Saturday from 9 a.m. to 12 p.m. The farm store is open 9 a.m. to 5 p.m. daily but call ahead.

Buttonwood Farm Ice Cream

473 Shetucket Turnpike, Griswold; 860-376-4081; buttonwoodfarmicecream.com

Buttonwood Farm has acres of sunflowers and a corn maze, but let's face it, most people go there for the ice cream. They churn out dozens of flavors, and the portions are huge. Traditionalists can try Coffee Mocha Crunch, while the more adventurous can sample frozen pudding flavors with rum, raisins, peaches, pineapple, maraschino cherries, and apples. Bring the kids in autumn for a run through the corn maze, or at other times of year to meet the cows and watch the sunflowers in the field turn toward the light as the day goes by. Dairy farming has become a specialty business in Connecticut, and Buttonwood has figured out the formula for success. Open March to November, 12 to 9 p.m. every day.

Cato Corner Farm

178 Cato Corner Rd., Colchester; 860-537-3884; catocornerfarm.com

Elizabeth MacAlister's amazing artisanal cheeses have won fans from across the country, and even across the ocean in France. Her cows are subjected to no hormones, herbicides, or chemical fertilizers—one reason their milk is so pure. But the other reasons are legion. The underground cave that Elizabeth built to keep the raw milk cheeses at the proper temperature might be one explanation. Their creamy Brigid's Abbey cheese and the popular Bloomsday are two varieties not to miss here. Also try the Womachego or the aged Jeremy River cheddar, both spectacular examples of their kinds. The farm shop is open Friday and Saturday 10 a.m. to 5 p.m. and Sunday 11 a.m. to 5 p.m., but you'll also find their creamy products at farmers markets and stores around the state.

Deerfield Farm

337 Parmelee Hill Rd., Durham; 860-301-7828; deerfieldfarm.org

The herd of registered Jerseys at Deerfield are raised humanely, without antibiotics or synthetic hormones. That means the raw milk, yogurt, chocolate milk, and soft cheeses they produce will be as healthy as possible. Their fencing allows the cows to rotate through

small fields as they crop the grass, making sure the land is not overgrazed or abused. During the dry months, they feed the cows hay grown on their land with no chemical pesticides or insecticides added. "Real cows eat grass," say the folks at Deerfield. Tasting their milk will have you supporting that argument wholeheartedly. Stop by from 10 a.m. to 6 p.m. daily to sample some. They raise pork and poultry in the same way, so pick up some bacon and eggs while you are there.

The Farmer's Cow

49 Chappell Rd., Lebanon; 860-642-4600; thefarmerscow.com

The Farmer's Cow is a group of six Connecticut family-owned dairy farms, including Cushman, Fairvue, Fort Hill, Graywall, Hytone, and Mapleleaf Farms. You can visit these farms individually and take tours. They specialize in dairy products, of course, but also sell locally sourced eggs, coffee, and apple cider. Their milk can be found in grocery fridges throughout the state, and they feature a home delivery service, sponsor a farm maze, and showcase educational farm tours. The jewel in their crown might be the Farmer's Cow Calfé & Creamery, at 86 Storrs Road, Mansfield Center, open Monday through Saturday 9 a.m. to 9 p.m. and Sunday 9 a.m. to 6 p.m. There you can get breakfast, lunch, and dinner, all washed down with their delicious milk or a local Hosmer Soda.

Sweet Grass Creamery

51 Mattern Rd., Preston; 860-887-8098; sweetgrass-creamery.com

With 100 percent Jersey cows, Preston's Sweet Grass Creamery is known for high levels of protein and calcium, but also for its flavorful cream-line milk, small-batch yogurt, and sweet cream cheese. With 200 acres for the cows roaming, milk production is secondary to the comfort and health of the cows. It's also a great destination to learn how this vital part of our diet is produced and made. Visit their market store for their freshest products, as well as many other local delights from neighboring farms. Open Tuesday to Sunday 10 a.m. to 12 p.m. and again from 2 to 5 p.m. "It's all about the cows," say the Sweet Grass crew, and they should know.

Terra Firma Farm

564 Norwich Westerly Rd., North Stonington; 860-535-8171; terrafirmafarm.org

Directed by Brianne Casadei (Farmer Brie), Terra Firma is a non-profit community farm and creamery, but also so much more. With the small herd of grass-fed Jerseys and a state licensed bottling room, the farm makes pasteurized whole milk, chocolate milk, coffee milk, yogurt, and soft cheeses. You will find a beautiful layer of cream on the top of their milk, so make sure you shake the bottles before serving. However, like many smart farmers in Connecticut, Terra Firma has diversified, with 100 percent pasture-fed beef, natural-habitat heritage breed pork, open-space broiler chickens, layer chickens, specialty poultry, and rabbits. Their farm store also sells prepared dinners, baked goods, soups, and more from a local restaurant, La Belle Aurore, and a selection of other products from nearby farms and craftspeople. They also handcraft their own Faire Ivy Soaps, with herbal extracts for all skin types, all grown in the summer gardens. Their farm store is open Wednesday through Friday 10 a.m. to 6 p.m., Saturday and Sunday 10 a.m. to 2 p.m., and Monday 3 to 6 p.m.

Committed to true community agriculture, Terra Firma does more than that, opening for educational field trips, programs, and various camps that reconnect children to agriculture. If you live within a 30-mile radius, they will deliver to your home once a week. And in 2019, they launched the Terra Firma Farm Give Gallons campaign, a milk purchase program that delivers to the Pawcatuck Neighborhood Center and the Gemma E Moran United Way/Food Center. Now that's true dedication.

Farms and Farm Stands

Campbell's Farm Stand

1 Campbell Rd., Jewett City; 860-376-1066; campbellsfarmstand.com

This seventh-generation family farm sells their farm raised, all natural, USDA inspected beef at their seasonal farm stand. These cattle are raised safely, humanely, and hormone free, fed with corn from the farm and grains from a local mill. You'll find steaks, ground beef, patties, hot dogs, kielbasa, and more. Their farm stand also sells numerous vegetables, from sweet corn to squash, asparagus to rhubarb. Jam, honey, salsa, and freshly made pies are also featured.

Cold Spring Farm

46 Lake Hayward Town Rd., Colchester; coldspringfarmct.com

This year-round, full-diet farm sits on 300 acres above beautiful Lake Hayward and is a tribute to how diversification helps Connecticut farmers survive. The list of vegetables, fruit, and herbs that grow on this farm is truly staggering, with over 12 kinds of lettuces and greens alone. They have fields of edible flowers and more herbs than you could ever possibly use in the kitchen. Pasture-raised meats, canned goods, and more at their farm stand, open every day of the week, every day of the year. You can also get their food through a CSA or at several regional farmers markets. Or, if you don't mind a little hard work, you can put in daily labor at the farm for a weekly share of fresh veggies, flowers, fruits, and herbs. "If we use what nature provides to build our soil health and grow our food," say farmers Jess and Jeff, "the results are astounding and best of all, delicious."

Fork It Farm

174 Clark Hill Rd., East Haddam; 860-661-2042; forkitfarm.com

The Newton family of Fork It Farm has a passion for health and sustainability. Danielle Newton has wanted to be an organic vegetable farmer since she was a teenager, and after marrying carpenter Greg Newton, they were able to make the dream come true. Ground cherries, tomatoes, garlic, chard, and spicy peppers are all available at their small farm stand or by free local delivery. The stand is open 8 a.m. to 8 p.m. daily, usually operating by self-service. Danielle also offers yoga and kirtan classes on the farm, so check their website and Facebook page for events.

Four Mile River Farm

124 Four Mile River Rd., Old Lyme; 860-468-4040; fourmileriverfarm.com

Since 1985 the Corsino family farm has bred Jersey cattle on acres of lush pasture. Their beef is top notch, with many different cuts available. They also raise pigs in open-air shelters and feed them only high-quality grain. The pasture-raised chickens use mobile coops, and you'll find both chicken and farm-fresh eggs for sale. We enjoyed their sausages for a 4th of July barbecue, and they were some of the best we ever tasted. Their year-round farm stand is open Monday through Saturday 8 a.m. to 7 p.m. and Sunday 8 a.m. to 6 p.m. "Food equals family," say Nunzio and Irene Corsino. You'll find that connection in full force at Four Mile River.

Halfinger Farms

489 Candlewood Hill Rd., Higganum; 860-345-4609; halfingerfarms.com

Surrounded by rolling hills, the Halfinger family farm is a hidden gem. From March to May a pick-your-own daffodil farm operates on this historic property, and they have a greenhouse with pansies and perennials (among other plants). In late May the vegetables start arriving, with herbs and 30 varieties of tomatoes. Unusually, they

close for retail sales from mid-June until September, when they re-open to sell thousands of colorful Belgium mums. In autumn Halfinger also features pick-your-own pumpkins and a corn maze. Their spring hours are usually Monday through Friday 9 a.m. to 6 p.m. and Saturday and Sunday 8 a.m. to 5 p.m. In September and October, they are open for the same hours.

Holmberg Orchards

12 Orchard Ln., Gales Ferry; 860-464-7305; holmbergorchards.com

Established in 1896, Holmberg's farm and orchards include seasonally available pick-your-own fruits like strawberries, blueberries, and raspberries, and the orchard trees bear apples, peaches, nectarines, and pears. Stop at the farm market located just below the farm on Route 12 for fresh pies, muffins, and pastries, as well as a colorful array of farm-fresh fruits and vegetables, including lavender in the summer and pumpkins in the autumn. You'll also find wine from local wineries, as well as Holmberg's own fruit wines and ciders. Fourth-generation Russell guides the winery and cider press, making delightful Pinot Blanc and full-bodied, tart hard cider with russet apples. Friday through Sunday 12 to 6 p.m., sample the wine and tour the orchards and grapevines. Look out for curiously placed glass bottles hanging from

COURTESY OF LEE EDWAR

COURTESY OF LEE EDWAR

pear branches. These are secured around blossoms as they bud, and the pears grow in the glass. Holmberg partners with Connecticut's Westford Hill Distillery, which fills the pear-trapped bottles with its distilled *eau-de-vie*. Pick-your-own fruit mornings 8 a.m. to 12 p.m. and look for special events like the "wine maze" in the fall.

Lavender Pond Farm

318 Roast Meat Hill Rd., Killingworth; 203-350-0367; lavenderpondfarm.com

Walking the 25 acres of Lavender Pond Farm you'll find 10,000 lavender plants, but also a profound sense of the beauty of the world. This unique, solar-powered farm in Killingworth features a purple train (really a sort of awesome tour bus), and they give educational tours every hour on the hour during the high season. You can check out their copper alembic still, from which they make lavender oils, and play giant chess in the formal garden. Starting June 2, the hours are 10 a.m. to 4 p.m. rain or shine., and the lavender is blooming through July. If you go at the right time of day, you will see the thousands of bees that this farm is helping to nurture. Those bees mean the future for many types of agriculture, and maybe the future for human life on this planet.

Maple Lane Farms

57 NW Corner Rd., Preston; 860-889-3766; maplelane.com

With a diversely planted 325 acres, Maple Lane Farms could fit anywhere in this book. In November and December, you can cut your own Christmas tree from their selection of Balsam Fir, Fraser Fir, Canaan Fir, and Concolor Fir seven days a week. Approximately 250,000 heads of their hydroponic Bibb lettuce grow year-round in the greenhouse. They also offer a farm stay in the 1791 house through airbnb.com.

Millington Beefalo Farm

57 Millington Rd., East Haddam; 860-608-8558; millingtonbeefalo.com

What, you haven't heard of beefalo? Well, it's exactly what you'd expect—a cross between North American bison and domestic cows.

The meat from these hybrid wonders has less cholesterol, fat, and calories than regular beef or buffalo, but is higher in protein and minerals and is the only red meat shown to help lower LDL cholesterol. The farm-raised, grass-fed beefalo at Millington are given no antibiotics, steroids, or hormones. You can visit the farm and pick up some of this excellent meat on Saturday and Sunday 10 a.m. to 4 p.m.

Penfield Farm

51 Ames Hollow Rd., Portland; 860-836-5599; penfieldfarm.com

This farm has been family owned since the early 1800s and today humanely raises hormone- and antibiotic-free animals in a grass-fed, free-range environment. The farm store sells aged black Angus beef, Berkshire pork, lamb, chicken, duck, and even emu! You'll also find farm-fresh eggs with high levels of omega-3 fatty acids, super-healthy dog food, and Songline emu oil products. They also have educational opportunities for school kids, who love to interact with the many rescue animals living in their forever home on the farm. The store is open every Saturday 9 a.m. to 3 p.m. and at other times by appointment.

Provider Farm

30 Woodbridge Rd., Salem; 860-705-3673; providerfarm.com

Farmer Hannah has created a small wonder at Provider Farm. Their CSA is 23 weeks long beginning in June and full of sustainably grown produce at a reduced price. They also offer a pick-up site on the farm, which allows you to order by Thursday evening and get your food (including meat, eggs, dairy, mushrooms, and all their veggies) on Saturday at the farm shop. Or you can stop by on Saturday from 1 to 4 p.m. to browse their selections, along with other delights like bread from Nana's Bakery and candles from Bright Edge Candles.

Sankow Beaver Brook Farm

139 Beaver Brook Rd., Lyme; 860-434-2843; beaverbrookfarm.com

The Sankow family is blessed with over 700 Romney sheep that live on the farm and graciously donate their wool to make blankets,

COURTESY OF LEE EDWARDS

hats, scarves, sweaters, and mittens. However, the farm also features cows and East Friesian sheep, which they milk to make cheeses, like Nehantic Abbey and a nice feta. They also have a thick farmstead yogurt that will completely change your ideas about this too often mass-produced breakfast food. At their farm market, open 9 a.m. to 4 p.m. seven days a week all year, you can also buy chops and sausages, vegetables, shepherd's pie, and ice cream. Sankow's cheese is considered some of the best in the US and is at the forefront of the new movement that will hopefully put America back on par with the great cheese makers of Europe. Like many Connecticut farmers, they wear many hats. Theirs happen to be wool.

COURTESY OF LEE EDWARDS

Stone Acres Farm

393 North Main St., Stonington; 860-245-4414; stoneacresfarm.com

This 63-acre working farm in gorgeous Stonington opens its rolling hills and gardens to visitors every year in many different ways. Their farm stand is open Monday through Friday 10 a.m. to 6 p.m. and 9 a.m. to 5 p.m. on the weekend, and you can find a wide variety of local vegetables and greens, eggs and dairy, meat and honey. Wood-fired bread and artisanal cheeses make this the perfect place to pick up a picnic lunch. However, that is only one of many attractions at Stone Acres. Two greenhouses dating from the early 1800s produce fresh citrus, perennials, and grapes year-round. The farm hosts farm dinners, culinary workshops, and other events throughout the year, and their Yellow Farmhouse Education Center preserves agricultural heritage through a variety of programs for teachers, school groups, and others. The farm also features a 12-week CSA program for those in the area. One of the most special products is a Farmer's Choice Bucket, with 6–8 bunches of flowers at once, available May through October.

West Green Farm Orchard and Dairy

119 West Town St., Lebanon; 860-642-6745; westgreenfarm.com

This former 32-acre dairy farm was purchased by the Preli family in 1998, and though they keep cows, they have diversified a lot. Six acres of fruit, 3 acres of vegetables, and herbs, flowers, and honey make for a more balanced farm, and a more balanced diet. You can buy their produce from the farm stand, pick-your-own during the season, or join their 16-week fruit and veggie CSA. The cows produce raw milk, which you can get weekly supplies in a separate CSA. The farm store also sells fiber and yarn crafts from local artisans and sponsors knitting classes. It is open June through October, every day 6 a.m. to 8 p.m.; in November and December, Thursday through Sunday 12 to 4 p.m.

The store is in an 18th-century barn and is a couple doors down from Governor Jonathan Trumbull's "War Office," really the family's farm store. From that small building, Trumbull and his family fed George Washington's entire northern army and helped win the Revolutionary War. Your needs might not be that extreme, but whatever they are, you'll find them met at West Green Farm.

Whittle's Willow Spring Farm

1030 Noank Ledyard Rd., Mystic; 860-536-3083; facebook.com/Whittles-Willow-Spring-Farm

This small family farm has managed the economic crises of the past century with aplomb, and every summer locals stop by their charming store to pick up local fruits and veggies, eggs and honey. In the autumn they feature a harvest patch with fall mums, pick-your-own pumpkins, and pick-your-own McIntosh and Cortland apples at discounted rates. Usually open the first weekend in July through the

fall, the hours are Monday through Saturday 9 a.m. to 6 p.m. and Sunday 9 a.m. to 5 p.m.

Farm Breweries and Cideries

B. F. Clyde's Cider Mill

129 N. Stonington Rd., Old Mystic; 860-536-3354; clydescidermill.com

Established in 1881 and operated by the sixth generation of the Clyde family, this is the oldest steam-powered cider mill in the United States and a National Historic Mechanical Engineering Landmark. Open seven days a week September until late December 9 a.m. to 5 p.m., the cider mill also features cider-making demonstrations on Saturday and Sunday. But you can come any day in season to get their hard or sweet ciders and apple wines, pies, jams, and jellies, along with local honey, maple syrup, fudge, pumpkin bread,

COURTESY OF LEE EDWAR

gourds, Indian corn, pumpkins, candy apples, and kettle corn. Their apple cider donuts are top notch. But it's a treat just to be there, watching steam rising from the roof of the mill, hearing apples tumbling off the trucks, and smelling that sweet smell of fermentation.

Fox Farm Brewery

62 Music Vale Rd., Salem; foxfarmbeer.com

At this rural farm, complete with silo, red barn, and tasting room, you can sample some of the most interesting beers in Connecticut. Along with a selection of India pale ales, you'll find a foeder-aged farmhouse ale, Xenia honey beer, and a coffee stout made with locally roasted Ashlawn Farm coffee. Specialties include Susurrus, an ale with sumac, yarrow, and boysenberries; and Nettles, a farmhouse ale with, yes, nettles, as well as ginger and yuzu. Fox Farm has recently (during the pandemic) instituted a reservation system for anyone who wants to enjoy a tasting on-site, so get onto their website and make one. The tasting rooms hours are Thursday and Friday 12 to 7 p.m., Saturday 11 a.m. to 6 p.m., and Sunday 11 a.m. to 5 p.m. If you want to take home their beer in cans, you'll find their design strangely familiar—it was inspired by seed packaging.

Hop Culture Farms

144 Cato Corner Rd., Colchester; hopculturefarms.com

This hop farm and brewery on Cato Corner Road is family owned and family friendly. Their sharp tag line is "great beer grows here," and that statement is both an advertisement and a deep piece of wisdom about beer itself. Beer is an agricultural product, something we too often forget. Their newly expanded production facilities have led the brewers to such delights as Moon Boots Stout, Dad Bod IPA, and Blue Jean Baby, made with blueberries from nearby Savitsky Farm. Those who prefer a crisp, light beer should try the Think Piece Pilsner, while those who yearn for a fruity, sour beer should go for What the Fruit. Seasonal beers like Elf Yourself, a sweet potato stout, are on rotation, so check their website for current offerings. Check also for their schedule of food trucks, local music, trivia nights, and other special events, including ones that you can bring the whole family to. Open Thursday and Friday 5 to 9 p.m., Saturday 12 to 8 p.m., and Sunday 12 to 6 p.m.

Farmers Markets

Chester Sunday Market

23 Ridge Rd., Chester; 860-790-5007; chestersundaymarket.jimdo.com

More than just a farmers market, this is a townwide event that happens every Sunday from June to October 10 a.m. to 1 p.m., usually centered around the elementary school, but check the website for current location. Local vendors sell fruits and vegetables, cheese and breads, honey and flowers. Most years, the entire main street is closed off and music plays as the people of the town and surrounding communities stroll the streets, sipping coffee and perusing art. Events like this one do not just bring a town together, they create a town as a living entity, one where the people have actual neighbors, one that connects us to one another and to the land where we live.

Cromwell Farmers Market

1 River Rd., Cromwell; cromwellfarmersmarket@comcast.net; cromwellfarmersmarket.org

Cromwell's farmers market runs from early June through September, Friday from 4 to 7 p.m. This unusual evening market does all the usual things—provide customers with home-grown seasonal produce and crafts from local farms and artisans. But there is a reason that this market consistently ranks as one of the best in the state. That is because it is not just a market to buy things, it is a Friday night community event, with live music and food trucks. People bring blankets and eat dinner with their families and friends and dance the night away. Join them.

Durham Farmers Market

30 Town House Rd, Durham; townofdurhamct.org/farmers-market

On the village green Thursday 3 to 6:30 p.m. April to November, the Durham Farmers Market offers CT Grown and Connecticut-made products from two dozen vendors. You'll find maple syrup, cold-pressed juices, seasonal fruit, cut flowers, raw honey, goat cheeses, pickles, jams, handmade pasta, scones, hummus, kettle corn, candles, chips, and so much more. You'll also find a food truck or two, with empanadas, hot dogs, and lobster rolls. They also have a winter market, located at the Durham Activity Center (350 Main Street) Saturday 10 a.m. to 1 p.m.

Ledyard Farmers Market

728 Colonel Ledyard Hwy.; market_mgr@ledyardct.org; ledyardfarmersmarket.org

Since 2008, the Ledyard Farmers Market on the town green has been at the forefront of agritourism, creating special moments and experiences for visitors. In addition to being a certified CT Grown market with a curated set of vendors, Ledyard is a true community

COURTESY OF LEE EDWARDS

enterprise. Every year it funds itself with new market bags, and people flock to get them. That is a tribute to how much locals care about this special weekly event. As an evening market, they bring food trucks for dinners and brews to wash them down. Musical guests keep you entertained while you shop and gather with friends. Visit this impressive market on Wednesday from June 2 through September 15 from 4 p.m. to 7 p.m. Each week at the Ledyard Farmers Market is a different experience, with a new theme and new opportunities, so check their website and decide which week to visit . . . or just try all of them.

Niantic Farmers Market

Methodist St. (summer market) 4 West St. (winter market), Niantic; nianticmainstreet@sbcglobal.net; facebook.com/NianticFarmersMarket

This small market in the delightful village of Niantic is a great example of how the Connecticut Department of Agriculture is helping organizations like Niantic Main Street fund agritourism events. Matching funds from the Community Investment Act allow this market to operate on Thursday evenings from June 24 to October 14, 3 to 6 p.m. They bring in food trucks, author signings, meditation sessions, and more during the summer and winter at the Methodist Street parking lot. You will find baked goods and local meats, as well as other fun choices like a raw bar and Middle Eastern cuisine.

COURTESY OF LEE EDWARDS

Old Saybrook Farmers Market

210 Main St., Old Saybrook; oldsaybrookfarmersmarket.com

For 25 years, the Old Saybrook farmers market has been one of the most consistent and lauded in the state. Open late June to the end of October, usually Wednesday and Saturday 9 a.m. to 12:30 p.m., the market is held on the property of William Childress. There are usually about 30 vendors, all selling certified Connecticut products. You'll get meat from Stonington, pottery from Madison, and berries from Norwich, along with fresh-baked bread, kettle corn made on-site, and a variety of handcrafted items for sale. There is usually live music or other demonstrations, anything from martial arts to craft skills.

Stonington Village Farmers Market

22 Bayview Ave., Stonington; sviastonington.org/farmers-market

Begun in 1997, this market operates year-round at the Velvet Mill shopping center, inside during the winter, Saturday 10 a.m. to 1 p.m., and outside during the summer, Saturday 9 a.m. to 12 p.m. It brings in people from the surrounding towns, because a weekend in one of the most beautiful villages in America is always a treat. Vendors line up to sell locally grown vegetables and fruits, homemade breads and pies, cheese and eggs, poultry and seafood. Even in winter you can sample Italian ices and buy fresh-cut flowers. It's also a great place to pick up a picnic lunch (with picnic tables on the premises), with prepared food like soups, meat pies, and breakfast sandwiches.

Buy some merchandise from the Stonington Village Improvement Association while you're there, because it helps to support the continuation of this local treasure.

Farm Stays

Graywall Farms

724 Exeter Rd., Lebanon; 860-617-2262; thehillatgraywallfarms.com

Graywall is one of the Farmer's Cow cooperative farms, so probably you have already tasted milk from their cows. At 900 acres, it is one of the larger farms in this guide, although it is split between the dairy farm, protected land, and a horse boarding and training facility (The Hill at Graywall Farms). Nearby is a beautiful 1800s Federal-style home that is available to book for lodging. The Hill offers a number of training and experience packages for those who have their own horses and want some intensive education. They are also planning wine and cheese weekends, with expert cheesemaking classes in conjunction with the dairy farm part of the operation. Two nights lodging, a cheesemaking clinic, and a pass to local vineyards sounds like a great weekend to us. How about you?

Roseledge Farm Bed and Breakfast

418 Rte. 164, Preston; 860-892-4739; roseledge.com

This 1720 farmhouse has fireplaces in each of its three guest rooms. The innkeepers will invite you to help with the morning chores, and you should take this opportunity to feed the goats and sheep or collect the morning's eggs. There's even a beehive on the property, but don't worry, you won't have to smoke it to check the honey. Instead, you'll be treated to feather beds, fresh flowers, wood-burning fireplaces, homemade soaps, and evening wine. In addition to selling plants, garden products, and books, Roseledge's farm stand is also open to the public for tea and baked goods. You'll enjoy those in the morning of course, with eggs gathered daily—one of many reasons to regret ever having to leave this wonderful place.

Festivals and Fairs

Durham Fair

Durham Green, Durham; 860-349-9495; durhamfair.com

Begun in 1916, this is the largest agricultural fair in Connecticut. It takes place on the last full weekend in September on the Durham town green and the adjoining fields. You can visit the Fair Farm Museum, with antique equipment and collectibles, and the Discovery Center, with seminars on food, farming, and garden care. You'll be astonished at the animal pulls, where horses, oxen, and ponies compete the way they have for hundreds of years. Check out exhibits of livestock—unbelievable premium cattle, goats, llamas, poultry, rabbits, pigs, and sheep—and products, like baked and canned goods, crafts, needlework, photography, plants, flowers, fruit, and, of course, giant pumpkins. Every year about 12,000 entries compete for blue ribbons and a best-in-show rosette. There's a talent show for singers, dancers, and entertainers on the last day, and for the kids there are pie-eating contests and fun things like bubble stations. If these traditional fair events are not enough, there's also a demolition derby, a truck pull, a carousel, midway rides, and more. And don't forget the main stage, which has a long tradition of very impressive acts, from the Carter Family to Kenny Rogers.

Haddam Neck Fair

26 Quarry Rd., Haddam; 860-267-5922;
haddamneckfair.com/attractions

First organized in 1909 by the local grange, at a time when most people in the area were farmers, the Haddam Neck Fair still takes place around the Grange Hall built in 1911. However, it has grown considerably since those early days, with new attractions and entertainment added every year. Helicopter rides, dog agility contests, and the 5K Road Race draw visitors these days, but horse, oxen, and tractor pulls are still favorites. Women can still join the skillet-throwing competition, and men can enter the beard contest. And, of course, you can participate in or just browse through dozens of shows, from goats to sunflowers. Usually held on Labor Day weekend, with gates opening Friday at 4 p.m. and the other three days at 7:30 a.m.

Lebanon County Fair

122 Mack Rd., Lebanon; lebanoncountryfair.org

Usually held the weekend of the second Sunday of August, the Annual Lebanon Country Fair is sponsored by the Lebanon Lions Club. The food at this fair is top notch, with a combination of local groups and commercial vendors. The exhibition hall is filled with the pride and joy of area residents: baked goods and flowers, paintings and quilts. You can enter the horseshoe tournament or one of the state contests if you are the competitive type, or just enjoy the musical entertainment and rides if you prefer to relax. Kids will learn about the fine art of animal husbandry and adults will learn how to tell their children "no" when they ask for another plate of fried dough. It is open all three days of the weekend, but Saturday is the "long" day, from 8:30 a.m. to 11 p.m. if you want to see absolutely everything. Bring cash because the admission desk and some of the vendors do not take credit cards.

North Stonington Fair

21 Wyassup Rd., North Stonington; 860-535-3956; northstoningtonfair.org

Sponsored by the North Stonington Community Grange and the North Stonington Volunteer Fire Department, this fair is four days of fun every July. Start on Thursday evening with tractor pulls, amusement rides, and a band. On late Friday afternoon you can watch the cows come in with the dairy show, a baled hay toss, and trained Labrador Retrievers doing their thing. Saturday has an extended schedule, with all the animal shows, a kids' parade, a pie-eating contest, and wrestling. Yes, wrestling. Eat their famous ham and bean supper and come back on Sunday for the antique cars. Farm life never seemed so fun.

Maple Sugarhouses and Apiaries

Goshen Hill Maples

1040 Goshen Hill Rd. Lebanon, CT; 860-465-6582; facebook.com/goshenhillmaples

One of the newer players in Connecticut's maple syrup game, this post-and-beam sugarhouse on 14 acres already produces enough so you can find their products in retail stores across the southeast portion of the state. However, it's always the most fun to go to the sugarhouse itself. You can visit and watch the process mid-February to late March and buy year-round. Call ahead to make sure the boiler is going or if you are visiting in the off-season. Goshen Hill also participates in Connecticut's annual open sugarhouse weekend, and that's the best time to catch them doing a "live boil."

Stonewall Apiary

120 Inland Rd., Baltic; 860-964-3532; ct-honey.com

This family-owned apiary based in Hanover has over 350 honeybee colonies throughout eastern Connecticut. You'll find their honey, creamed honey, comb honey, and chunk honey in farm stores around the state, but also at their location in Baltic. It is difficult to find with a GPS (a fact they note on their website), so maybe write down the directions beforehand. You can visit their shop every Saturday morning 9 a.m. to 12 p.m. They give beekeeping classes for those who want to learn about this important practice, as well as presentations at schools and libraries around the state, complete with child-sized bee suit.

Sugar Maple Farm

779 Exeter Rd., Lebanon; 860-642-9984; sugarmaplefarms.org

Connecticut is tenth in maple syrup production in the United States, and small producers like Sugar Maple are the backbone of that industry. For over 30 years, this farm has set out thousands of lines and over 300 buckets every year, tapping trees in Lebanon and boiling the sap into thick, dark syrup. Give them a call for hours, with the obvious time to go being the maple season in February or March. In addition to various kinds of syrup and honey (try the honeycomb!), they often have fresh horseradish and duck and chicken eggs for sale. "We sell the best and eat the rest," is Sugar Maple's motto. It's up to us to make sure there are no leftovers.

Three Sisters Farms

11 Evans Ln., Essex; 860-575-5999; threesistersfarm.com

Just outside of one of the most beautiful villages in America, Three Sisters Farm sells raw wildflower and infused lavender honeys from their hives on both sides of the lower Connecticut River. Named for the Stasia and Glenn Penkoff-Lidbeck family's three daughters, this family business makes handcrafted soaps, beeswax candles, skin creams, and lip balms. Everything they make is purely natural, with no artificial scents, preservatives, colors, or petroleum-based products. Call ahead for hours of operation or find their products at the Old Saybrook Farmers Market and beyond.

Museums and Education

Zagray Farm Museum

544 Amston Rd., Colchester; zagrayfarmmuseum.org

This nonprofit educational farm museum in Colchester is operated by the Quinebaug Valley Engineers Association Inc. Engineers may at first seem like an unlikely group to be preserving our agricultural history, but you just have to think about all the farm equipment machinery to realize your mistake. This incredible collection of antique engines, tractors, and other machinery is set on 200 acres of farmland and woodland, with original buildings that include the Zagray family homestead, dairy barn, machine shop, foundry, and antique sawmill. The museum is always growing, with a giant new building planned. In the meantime, you can attend one of their three weekend-long shows every year and marvel at the machinery of the past.

Pick Your Own

Gotta's Farm

661 Glastonbury Turnpike, Portland; 860-342-1844; gottasfarm.com

Pick-your-own strawberries, apples, peaches, and pears is only one part of this marvelous farm in Portland, which perches on the ridges above the Connecticut River meadows. A fourth-generation farm founded in 1898, Gotta's also has a farm stand that sells their fruit, of course, but also their own vegetables, baked goods (pies, breads, and cookies!), and quality plants from their garden center. At Christmas time, shop for your trees and decorations here, too. You will also spot Gotta's at local farmers markets and at the QP Market on Route 66 in Portland (Monday through Friday 8 a.m. to 7 p.m. and Saturday and Sunday 8 a.m. to 6:30 p.m.; 860-342-5030). Their Glastonbury Turnpike farm stand and the accompanying fields are open Monday to Friday 8 a.m. to 6:30 p.m. and Saturday and Sunday 8 a.m. to 6 p.m.

Grant's Berry Patch

188 Mell Rd., Lisbon, CT; 860-376-5625; grantsberrypatch.net

Third-generation farmer Chris Grant and his wife, Michelle, own this delightful small farm just off Route 395 in Lisbon. For those of us who are a bit lazier, the farm stand is the place to go, but there are significant savings to be had if we're willing to pick your own vegetables and fruit, with no reservation or picking fee and almost half the price at the stand. Blueberries, strawberries, raspberries, and pumpkins are the main crops, but you will also find other summer and fall vegetables, including sweet corn and tomatoes. You can find the Grants at local farmers markets, and if you live nearby enroll in their CSA. When you go, remember that it is cash or check only. The farm stand is open Saturday and Sunday 9 a.m. to 4 p.m. and Monday through Friday 9 a.m. to 5:30 p.m. The pick-your-own is open from 8 a.m. to 2 p.m. but call ahead to make sure what fields are open.

Lyman Orchards

Rte. 147 and Rte. 157, Middlefield; 860-349-1793; lymanorchards.com

Since 1741 this family-owned farm has been a destination for generations of Connecticut families. There are pick-your-own apple and peach orchards, and pumpkin, raspberry, strawberry, and blueberry fields. The hugely popular Apple Barrel farm store features not just fresh produce and milk but also baked goods, cheese, fudge, and a small deli and cafe with wraps, coffee, and sandwiches. Lyman Orchards also runs seasonal festivals, such as a Strawberry Festival in June, with horse-drawn wagon rides, pony rides, pie-eating contests, and live music. You can subscribe to the CSA and even golf on their world-class courses.

Scott's Yankee Farmer

436 Boston Post Rd., East Lyme; 860-739-5209; scottsyankeefarmer.net

The Scott family has owned this farm in East Lyme for over a century, and it is occupied by current owners Tom and Karen Scott.

COURTESY OF LEE EDWARDS

You can visit the farm to pick raspberries, strawberries, blueberries, peaches, apples, pumpkins, and even flowers. Check their website for the schedule, but strawberries usually start on June 10 every year. The picking is slightly up the road at 444 Boston Post Road. There is often a sunflower festival in the fall and other events. They also have a cider mill to press their own apples, a cider donut-making room, and a greenhouse where you can find plants in the spring and pumpkins every fall. Tom and Karen have four children and six grandchildren, so we can probably anticipate that this farm will be a family farm for a long time to come.

Sea Farms and Aquaculture

J and R Scallops

22 Bayview Ave #42, Stonington; 860-415-4656; jrscallops.com

This seafood market run by former fishermen is more than just scallops, it is an aquafarm-to-table team. They sell to over 100 restaurants and other fish markets, but they also sell directly to us at their shop at the Velvet Mill in Stonington and at 45 Williams Avenue in Mystic. They also have home delivery for those within a 30-mile radius of Stonington, with a $50 minimum. The Velvet Mill is open Thursday and Friday from 3 to 7 p.m., Saturday 9 a.m. to 2 p.m., and Sunday 11 a.m. to 4 p.m., while the Mystic location is open every day but Tuesday, 10 a.m. to 6 p.m.

Mystic Oysters

100 Main St., Noank; 860-333-7961; mysticoysters.com

At the Noank Aquaculture Cooperative you'll find the market for Mystic Oysters year-round, Friday 12 to 4 p.m. and Saturday 10 a.m. to 2 p.m. They sometimes have clams, but what they specialize in is oysters. President Jim Markow started his career at the famous Blue Point Oyster Company and then started this business with Karen

Rivera in 1994 as a hatchery in the cellar of his house. In the 2000s they moved to Noank and harvest the cultured Mystic Oysters and the wild Ram Island Oysters. These two bivalves are grown a half a mile from each other but look and taste different. Stop by Mystic Oysters and decide which is your favorite.

Niantic Bay Shellfish Farm

111 Main St., Niantic; 860-739-6273; nianticbayshellfishfarm.com

Niantic Bay Shellfish Farm is committed to restoring the scallop to our area. By farming shellfish, they are creating not just a delicious treat for landlubbers like us, but a cleaner, healthier Long Island Sound. For example, one adult oyster filters 50 gallons of water per day. By hanging racks and cages in the vertical space of the water, Niantic Bay (and other aquaculture farms) create ecosystems teeming with life. Macroalgae and phytoplankton are consumed. This process is right now restoring the sound to what it once was—one of the most important offshore farming areas in America. So, support this process by stopping by their market store, open on Friday 11 a.m. to 6 p.m. and Saturday 11 a.m. to 3 p.m. You can also join their CSA, attend B.Y.O.B. tastings in February and March, and reserve a spot on their boat to take private tours to view the "farm" out in the bay. "At heart, those who work in the aquaculture field are farmers," say the folks at Niantic Bay. That has always been the case, even if each succeeding generation has to prove it.

Trail Rides

Allegra Farm and Horse-Drawn Carriage and Sleigh Museum

Rte. 82 and Petticoat Ln., East Haddam; 860-873-9658; allegrafarm.com

Okay, if you've never gone on a New England sleigh ride in the winter, get to it. The perfect place to do it is Allegra Farm on Lake Hayward, the largest authentic "livery stable" in Connecticut. First of all, they have a huge collection of horse-drawn carriages and sleighs, most of which have been in films or television shows, from *Amistad* to *Sex in the City*. The horses and carriages you see at Mystic

Seaport are from Allegra too. They have open carriages or coaches enclosed with glass windows for warmer season rides, and open sleighs with blankets for going over the river and through the woods. Make a reservation and check the snow conditions before you go.

Tree Farms

Geer's Family Tree Farm

141 Norman Rd., Griswold; 860-376-5321; geersfamilytree.com

A family farm owned by the third generation, Geer's has been in the community for more than 60 years, providing sand, gravel, mulch, and compost to homeowners and contracting for building needs. But they are also one of the best places to go for Christmas trees in the area, with Fraser Firs, Douglas Firs, Concolor Firs, White Pines, White Spruces, and Blue Spruces. When the Christmas season is upon you, head here to walk the 100 acres of live trees to choose your favorite, or pick up a fresh-cut tree along with wreaths and decorations at the Tree Outlet. On weekends, food trucks gather, and it becomes a party at Geer's. They know what they are doing as far as festivities—they hold weddings on their overlook hill every summer.

Peaceful Hill Farm

118 Clark Hill Rd., East Hampton; 860-267-4341; peacefulhilltreefarm.com

With over 25 years on a preserved piece of land, Peaceful Hill is one of those idyllic places to cut your family's Christmas tree. Kids can also watch the Christmas train and shoppers can visit the craft barn to pick up wreaths and ornaments to accompany the 20 varieties of trees available. They provide saws and also will provide assistance if you like, shaking and baling the tree or helping you load it into your car. They will also pre-drill the tree if you have a pin stand. From the day after Thanksgiving through December 24, Peaceful Hill is open Wednesday to Friday 12 p.m. until "dark" and Saturday and Sunday 10 a.m. until "dark." Sip a cup of free hot cocoa as the snowflakes drift by, and you'll think you stepped into a Norman Rockwell painting.

Staehly Farms

278 Town St., East Haddam; 860-873-9774; staehlys.com

This farm and farm winery boasts nearly 100 acres in the Seven Sisters Hills, near Gillette Castle and Goodspeed Opera House. They produce their own wines and ciders from their own fruit, including a tomato wine, the only one in New England. However, at their well-stocked farm stand they also sell a range of farm goods, including a variety of plants, vegetables, pumpkins, and Christmas trees. The winery is open May 4 through November 4, Friday and Saturday 12 to 6 p.m. and Sunday 12 to 4 p.m. The Christmas trees are available from November 24 to December 23, Saturday and Sunday 12 to 4 p.m. Staehly hosts festivals and hayrides in the fall and other events throughout the year. You can also find their drinks around the corner at the Yankee Cider Barn at 23 Petticoat Lane in East Haddam.

Winterberry Farm

104 Parker Hill Rd. Ext., Killingworth; 860-663-2747; winterberryfarmkillingworth.com

Walking between the hay bales at Winterberry Farm discussing which tree to bring home this year has become a tradition in Middlesex County. From White Pines to Korean Firs, the choices seem endless, and as you take your tree cart and handsaw, you might wonder if two trees might not be the wiser option. Or three? When you have brought your tree back to the baling tables, pick up a wreath and a handmade walking stick in the barn. Cash or check only at Winterberry, the day after Thanksgiving to Christmas Eve, Saturday and Sunday 9 a.m. to 4:30 p.m., and Monday through Friday 1 to 4:30 p.m.

Wineries

Chamard Vineyard

115 Cow Hill Rd., Clinton; 860-664-0299; chamard.com

Ideally situated in proximity to Long Island Sound's beneficial growing effects, Chamard Vineyard's beautiful fieldstone winery, tasting room, and bistro is grand and intimate, rustic and cosmopolitan, with wood beams and a stone fireplace. The vineyard has 15 acres planted, and creates great wines like Chardonnay with a touch of honey on the finish and Cabernet Franc with notes of smoky berry and a little butterscotch. Enjoy a glass on the patio terrace and view the vines across a small pond with the fountain spouting in the center. Open Wednesday to Saturday 12 to 8 p.m. and Sunday until 5 p.m., the winery offers a prolific list of singer-songwriters, jazz, and bluegrass performers throughout the year. The vineyard is also a spectacular and popular setting for weddings.

Jonathan Edwards Winery

74 Chester Maine Rd., North Stonington; 860-535-0202; jedwardswinery.com

Owner and winemaker Jonathan Edwards developed his winery on 48 acres of land that had been farmed for centuries. The majestic white barn that sits atop the slope and overlooks the vines is the picturesque snapshot of what we imagine when we think of a winery. It's a distinctive destination for events and weddings, and the winery provides tasters with a unique opportunity. Edwards travels across the country each season to harvest grapes and make wine in the Napa Valley, and then ships the product back to Connecticut for customers to enjoy. These wines show off the Napa Valley appellation and are everything you'd expect from California wines. But the flavor of Connecticut is only a glass away. Chardonnay, Pinot Gris, Gewürztraminer, and Cabernet Franc are all grown on-site. The Napa reds are big and bold, made with grapes like Petite Sirah that just aren't suited to Connecticut's growing season. But Cabernet Franc is a standout among the Connecticut reds, and the estate-bottled Chardonnay may just beat out the Napa version in your taste test. The tasting room is open seven days a week from 11 a.m. to 5 p.m., unless there is a wedding event (usually weekends May to October), and then it closes at 4 p.m.

Lebanon Green Vineyards

589 Exeter Rd., Lebanon; 860-222-4644; lebanongreenvineyards.wordpress.com

In the center of Lebanon this small vineyard and tasting room is behind the 1778 home of David Trumbull, whose father and son were both governors of Connecticut. The winery has been built inside the old Milk Room and you can taste their eight wines Friday to Sunday 12 to 6 p.m. Don't forget to walk the crushed stone paths of the green, the largest in New England, and one of the most important staging areas for the Continental Army in the American Revolution.

Maugle Sierra Vineyards

825 Colonel Ledyard Hwy., Ledyard; 860-464-2987; mauglesierravineyards.com

Just south of Foxwoods Casino in the 1740 Ledyard House is a small winery run by physicist Paul Maugle. Here, less than 5 miles from the sea on a 100-foot hill, Maugle makes wine mostly from his prize St. Croix grapes. He uses ingenious methods like removing the acidic seeds to reduce the tannins and make the wines easier to drink. He cold ferments small batches, using St. Croix to make rosé as well. The tasting room and performance space is a popular stop and is open year-round for your enjoyment, Thursday 12 to 5 p.m., and Friday to Sunday 12 to 6 p.m. The vineyard also hosts festivals and wine dinners, so check the schedule. Maugle is a perfectionist, and he will probably tell you all about his methods when you stop by. "You want to taste the fruit, not the barrel," he says. At Maugle Sierra, you will.

Preston Ridge Vineyard

100 Miller Rd., Preston; 860-383-4278; prestonridgevineyard.com

Since 2008, Preston Ridge is set on 60 acres on the peak of a ridge (thus the name) that has incredible views across southeastern Connecticut. They grow and make Traminette, Riesling, Chardonnay, Vidal Blanc, and more. Our favorite is the estate bottled Cabernet Franc, one of several made completely from grapes grown on their land. Open Wednesday and Thursday 12 to 5 p.m., Friday 12 to 8 p.m., and Saturday and Sunday 11 a.m. to 4 p.m. Check the schedule for events in the evenings in the tasting room.

Priam Vineyards

11 Shailor Hill Rd., Colchester; 860-267-8520; priamvineyards.com

Owner Gloria Priam named this winery for her Hungarian grandfather, who worked a vineyard near Budapest. As New England's only solar-powered winery, it is a certified natural bird and wildlife habitat and the first vegan-certified vineyard in the state. On Friday in summer and fall you'll find a farmers market here; they also team up with a local chef and pair food and wine in a farm-to-table event. Like most Connecticut wineries, Priam creates original flavors that fit the climate and terroir. Their Riesling and Gewürztraminer bring German and Austrian winemakers here to buy cases for themselves. Open Wednesday to Sunday 11 a.m. to 6 p.m. May to September and Friday to Sunday the rest of the year. Along with tastings, there are musical events and food trucks, so check their calendar before you go.

Saltwater Farm Vineyard

349 Elm St., Stonington; 860-415-9072;
saltwaterfarmvineyard.com

Set on a peninsula in a bay next to Stonington Village, Saltwater Farm Vineyard could almost be listed under aquaculture. What is today a farm used to be an airport, and if you'd like to land your plane there, you still can (but no drinking and flying, please). The beautifully remodeled tasting room is in the old airplane hangar and can be booked as one of the most stunning banquet facilities in the state. But make reservations early; weddings are often booked years in advance. Sit on the porch overlooking the sound and sip the appley Sauvignon Blanc or rich, smoky Cabernet Franc. This is also a great spot for birding when the tasting room is open from April through November Sunday, Monday, Wednesday, and Thursday 11 a .m. to 5 p.m., and Friday and Saturday 11 a.m. to 3 p.m. Don't miss the osprey nests to your right as you drive in. Stroll the vineyard and watch kayakers paddle these secluded coves and salt marshes, or better yet, paddle them yourselves.

Stonington Vineyards

523 Taugwonk Rd., Stonington; 860-535-1222; stoningtonvineyards.com

With the ocean and its moderating breezes just miles away, Stonington Vineyards is ideal for grape growing. The vineyard has been in operation since the mid-1980s and specializes in Burgundian-style wines, styled by expert winemaker Mike McAndrew, who spent decades crafting the best wines in the state. Try two versions of Chardonnay—one fermented and barreled in oak, the other in steel—to experience why this grape is so versatile. The latter, Sheer Chardonnay, is comparable to the best versions of French Chablis, with mineral notes characteristic of the limestone terroir. Seaport White and Triad Rose are two delightful blends, and Stonington's version of Cabernet Franc shows why the coast's cooler temperatures, long growing season, and moderate winters suit this grape. The tasting room and facilities are housed in a long white structure, and the grounds are a popular choice for weddings. The winery is open year-round, seven days a week 11 a.m. to 5 p.m. and is easily accessible off I-95. Daily winery tours offer a glimpse of the state-of-the-art machinery and introduce visitors to the techniques behind their great wines.

APPENDIX A:

Organizations That Support Connecticut Farming

At its roots, agriculture thrives with community support—farmers helping farmers, consumers supporting local food growers, government organizations working to advocate for the good of all. It's hard work, made even more daunting in the age of consumerism, climate change, and economic uncertainty. Nevertheless, help is available, through a network of nonprofits, government organizations, and advocacy groups. Here is a list of resources for farmers, consumers, and activists.

Connecticut Agricultural Education Foundation

ctaef.org

The Connecticut Agricultural Education Foundation builds programs and acquires funds to fight agricultural illiteracy and promote Connecticut agriculture education. The organization coordinates Ag in the classroom events, funds a scholarship program, and coordinates Farm City events, a traveling multi-day workshop to bring area students to host farms to learn about farming and agriculture. Annual awards are given for Outstanding CT Farmer, Century Farm of the year, and AG Journalism, for outstanding news coverage of Connecticut agriculture. Awards are given at Ag Day in Hartford, usually the third Wednesday of March.

Connecticut Agricultural Experiment Station

ct.gov/caes

The Connecticut Agricultural Experiment Station has been working to improve agriculture in the state since 1875, by promoting scientific research, experimentation, and education. CAES plays a vital role in almost every aspect of the management of agriculture, supporting several departments including Environmental Science, Forestry and Horticulture, Plant Pathology and Ecology, and Entomology. CAES supports laboratories in New Haven and Windsor, the Griswold Research Center, and Lockwood Farm in Hamden.

Connecticut AGvocate Program

860-345-3977; agvocatect.org

The AGvocate program is managed by the Connecticut Resource Conservation and Development Area and partners with multiple state agencies to provide information and support for farmers, municipal governments, planning commissions, and residents. The program led to the creation of town commissions in the eastern part of the state, from Ashford to Woodstock. Resources for creating farm friendly communities include regulation ordinance guides, farmers market reference guides, templates to create townwide farm maps, and templates for conservation and development plans.

Connecticut Beekeepers Association

ctbees.org

The Connecticut Beekeepers Association supports the state's beekeepers and provides a forum for community involvement, education, and the advancement of scientific knowledge about the challenges and benefits of beekeeping. The group works to increase public awareness and promote healthy hives and sale of native honey and hive products. Their website offers resources for hobbyists as well as larger beekeeping enterprises including mentorships, workshops, hive registration procedures, and honey extractor rentals.

Connecticut Christmas Tree Growers Association

ctchristmastree.org

Did you know that the average Christmas tree takes on average seven years to grow to maturity? The CCTGA helps tree farms keep the holiday tradition going. First organized in 1960, the CCTGA supports the state's Christmas tree industry and helps consumers find, purchase, and care for the Christmas trees. Their objectives are to share information and resources about the production, research, and sale of quality trees, and to assist growers and buyers toward these goals.

Connecticut Department of Agriculture

860-713-2500; ct.gov/doag

The Department of Agriculture is the government agency in charge of developing, promoting, and regulating agriculture and aquaculture in the state through a network of programs, bureaus, grants, and nonprofits. The department oversees agricultural policy with the goal of advancing the economic viability of agriculture and related industries, protecting the state's resources and cultural heritage, and enforcing laws that secure the health of citizens and animals. The Department of Agriculture developed the CT Grown Program in 1986 (ctgrown.org), which helps the growth of agritourism with its green-and-blue logo that helps consumers find and support the state's agriculture and aquaculture industries.

Connecticut Farm Bureau

860-768-1100; cfba.org

Founded in 1915, the Connecticut Farm Bureau is the voice of Connecticut agriculture, advocating for farmers and promoting the importance and economic viability of farming in the state. The non-profit sponsors an eight-county network of over 3,000 members who support farmers and preserve the state's landscape through education, legislative advocacy, and market promotion. The bureau provides resources such as emergency preparedness education, food and farm safety links, farm labor and land acquisition information, and an advocacy toolkit.

CT Farm Energy Program

860-345-3977; ctfarmenergy.org

The Connecticut Farm Energy Program, developed and funded through the Connecticut Resource Conservation and Development organization (CT RC&D), provides grant assistance and loans to farms and small agricultural businesses. They aim to bring attention to energy conservation and help farms meet energy efficiency goals through renewable and alternative energy projects.

Connecticut Farmland Trust

ctfarmland.org

The Connecticut Farmland Trust works to preserve farmland and ensure a healthy environment for future generations. The trust does so through donations of agricultural conservation easements, direct purchase of farmland, and by allying with farmers, local officials, community organizations, and other conservation organizations to tackle farmland management concerns and ensure stewardship of the state's farmland. In over 20 years of operation, the CT Farmland Trust has protected more than 50 farms and nearly 4,000 acres of farmland.

Connecticut Farm-to-School Program

860-713-2503; ct.gov/DOAG/Farm-to-School/Farm-to-School/Welcome-to-the-Connecticut-Farm-to-School-Program

The Farm-to-School Program in Connecticut is a win-win for food producers, consumers, and communities, particularly for school-aged children who get access to healthy food produced by local farmers. The program, funded by the Department of Agriculture, aims to address how schools purchase food, and helps schools incorporate better practices and provide hands-on educational opportunities for students.

Connecticut Food Bank

ctfoodbank.org; foodshare.org

Part of the nationwide Feeding America network, the Connecticut Food Bank works with food growers and retailers to source healthy food for people in need. The organization works to collect and distribute food to the 490,000 people in the state who are food insecure and lack access to healthy foods. They network through mobile food share units, food pantries, soup kitchens, and support Senior Supplemental Food and Supplemental Nutrition Programs (SNAP).

Connecticut NOFA

ctnofa.org

Connecticut Northeast Organic Farming Association works to promote sustainable farming methods and develop economically viable opportunities for organic growers. Resources include organic certification consultation, soil health workshops, peer-to-peer mentoring, and networking opportunities.

Connecticut Resource Conservation and Development

860-345-3977; ctrcd.org

CT RC&D, a statewide nonprofit, works to ensure food security through sustainability project development and community outreach. The organization brings farmers, government entities, and citizens together for many conservation efforts, including infrastructure improvement, agritourism initiatives, and rural business growth.

Connecticut Sea Grant

seagrant.uconn.edu

Funded and administered by UCONN and the National Oceanic and Atmospheric Administration, Connecticut Sea Grant applies the mission goals of research, education, and outreach to the coastal and marine ecosystems.

Connecticut Vegetable and Berry Growers Alliance

cfba.org/ct-vegetable-berry-growers-alliance

Under the direction of the CT Farm Bureau, CVBGA provides members the opportunity to voice their ideas, prioritize issues specific vegetable and berry farming, and share best practices.

Maple Syrup Producers Association of CT

ctmaple.org

The Maple Syrup Producers Association of Connecticut (MSPAC) connects sugar makers of all levels to each other and to resources to start or continue successful commercial operations. Their mission is to help create high-quality maple products by recognizing the sugaring community, whether you're interested in introductory-level learning or visiting the network of sugarhouses around the state.

New Connecticut Farmers Alliance

newctfarmers.com

This statewide network aims to help new and emerging farms and encourages support for young and established farmers. The alliance provides support to ensure the success of the state's diverse agricultural community. Resources include access to peer-to-peer learning communities, planning and development guides, listservs, helpful information about land acquisition, taxes, growing and selling products, and energy efficiency methods.

University of Connecticut and UCONN Extension

cahnr.uconn.edu

The College of Agriculture, Health and Natural Resources at the University of Connecticut continues a mission started in 1881 to support the state's agricultural industry and prepare for a sustainable future. CAHNR provides opportunities for faculty, students, and the public to work together to build sustainable landscapes, tackle the challenges of climate change, and ensure the health and well-being of all. The extension network extends across all elements of the agricultural industry. The Storrs Agricultural Experiment Station works on research initiatives while UConn Extension focus on community outreach.

Other UCONN funded projects include the Center for Land Use Education (clear.uconn.edu), which assists with water management, land use planning, mapping, and STEM education. Farm Risk Management (ctfarmrisk.extension.uconn.edu) helps farmers and agribusiness reduce risk and increase financial know-how.

Working Lands Alliance

workinglandsalliance.org

Working Lands Alliance, a project of the American Farmland Trust, has been working to preserve farmland in the state since 1999. The statewide coalition drives policy through collaboration between farmers, lawmakers, and other stakeholders. WLA works to educate community leaders and sponsor advocacy initiatives. Their efforts have yielded the Community Investment Act, dedicated to conserving open space and preserving farmland and historic sites.

APPENDIX B:
Statewide Trails

When you experience the state through a particular lens, whether by visiting farms or going to museums, you inevitably see parts of the whole that you would otherwise miss. A map leads you in the direction of exploration, and the journey offers more than you asked for. The Connecticut Office of Tourism and their website *CTVisit .com* direct visitors to statewide events, attractions, arts and culture, and lodging; and offers interactive maps, guides, and vacation planning. In addition, there are organizations and passionate individuals who have mapped out common interests. Add a locally grown element to your next stay-cation and plan your own expedition. Here are some statewide trails to shape your itinerary.

Barns Trail

connecticutbarns.org/barns-trail

The weathered clapboard barn is an iconic symbol of farming in New England. Whether you're interested in historic preservation or seeing a working farm, the Connecticut Barn Trail offers something for everyone. The trail map and website link to seven self-guided tours organized by region. Barns are noted by architectural type—from the classic New England to the Polygonal—and by purpose—from corn crib to tobacco shed. Importantly, the trail map brings you close to other amenities, like pick-your-own, farm-fresh produce, and specialty shops. There's even an app. While on your barn journey, check out the New Milford Barn Quilt Trail, featuring 19 quilt patterns painted on the sides of vintage barns throughout New Milford (newmilfordfarmlandpres.org/barn-quilt-trail). These patterns pay homage to the long tradition of quilt making, and the trail offers an interactive way to explore the history and artistry of Connecticut's agricultural past.

Cheese Trail

buyctgrown.com/ct-cheese-trail

Connecticut farmers are raising cows, sheep, and goats to be happy and healthy, and happy animals make better cheese. The great thing about food lovers who are passionate about their home and local culture is that someone has mapped out the route from Deerfield Farm's chèvre to Cato Corner's underground cave. Listings compiled by websites like BuyCTGrown make it easy to map out your route and find farms as well as specialty cheese shops. Think beyond cheese, and indulge in farm-fresh milk, yogurt, and ice cream, and find your favorite.

Chocolate Trail

ctvisit.com/trail/chocolate

With over 20 chocolate specialty shops, with world-class chocolatiers and local artisans, the chocolate trail offers yet another unique way to explore the state and support the economy. Must-see destinations include Thorncrest Farm Milk House and Chocolates, where you can visit the barn then head to the store for milk, yogurt, and, of course, chocolate. At Munson's, the state's largest chocolate makers, they craft their chocolate from locally sourced milk and fair-trade suppliers. Trail listings are easy to navigate in list or map form.

Farmers Market Trail

farmersmarkettrail.com

The CT Farmers Market Trail showcases 15 markets across the state from Westport to Stafford Springs and over to Stonington. The website encourages you to buy local and support the state's network of markets with easy-to-download maps and an easy-to-use Google map link. You can even export the calendar of events and plan your week. Find them on Facebook for additional links and listings.

Ice Cream Trail

ctvisit.com/articles/sundae-drives-ice-cream-trail

No summer jaunt would be complete without an ice-cream cone and someone to share it with (or not). When you visit the farm where the milk was sourced and cream hand-churned, that's an added scoop of specialness. CTVisit has many listings that feature the state's best ice cream. Check out farm favorites and do your own taste tests. A few of ours include (clockwise southwest to southeast) Ferris Acres Creamery in Bethel, Arethusa Dairy Farm in Bantam, Farmer's Cow Calfé and Creamery in Willimantic, and Button Wood Farm in Griswold.

Live Local Trail

buyctgrown.com/uconn-trail

Brought to you by the intrepid people at BuyCTGrown.com and the UCONN Extension Center, the Live Local Trail links to several sites (and also has an app) connecting you to places around the University of Connecticut campus that support local agriculture. Attractions are open to anyone who wants to buy locally grown food and locally grown plants. To get to know the campus, visit the UConn Dairy Bar for some fresh ice cream, eat at farm-to-chef restaurants like Chuck and Augie's, and visit the animal science facility to learn more about the programs and check out the animals.

Organic Farm Trail

guide.ctnofa.org

Connecticut Northeast Organic Farming Association's Local Farm and Food Guide connects you to fresh food and farm products from organic farms across the state. Listings are organized by county to suit your day trip. Along with the directory, use the interactive map to find a site near you and new ones to explore. Nurture a stronger relationship with the land and food that you eat by championing organic growers and encouraging organic farming, and artisanal and locally made products.

Wine Trail

860-216-6439; ctwine.com

With over 30 wineries in the state, there are plenty of opportunities to experience varied landscapes and taste excellent wine. The Connecticut Vineyard and Winery Association, in partnership with the Connecticut Wine Development Council, sponsors the Wine Trail, along with the Passport Program. Visit a winery from May to

October and get a stamp at each tasting room and enter a drawing at the end of the season. The Wine Trail's brochures and website offer maps, hours, and sponsors trails signs and road markers to help you navigate from the Western Highlands to the Quiet Corner and along the coast. Taste a rich and earthy Cabernet Franc, then a lovely and floral Cayuga White, or your favorite Chardonnay fermented in steel, crisp and satisfying. Check out the calendar and plan your visit to hear live music and grab a lobster roll at a food truck on-site.

INDEX

G

H

I

P

R

S

ABOUT THE AUTHORS

PHOTO BY DAVID K. LEFF

Eric D. Lehman and **Amy Nawrocki** have co-authored several books about Connecticut, including *A History of Connecticut Food* and *A History of Connecticut Wine*, and both write for *The Wayfarer*, *Estuary*, and *Edible Nutmeg*. Eric is the author or editor of 21 books of travel, fiction, and nonfiction, including *New England Nature*, *9 Lupine Road*, and *Yankee's New England Adventures*. Amy is a poet and author of 10 books, including *Four Blue Eggs*, *The Comet's Tail: A Memoir of No Memory*, and *Mouthbrooders*. They are married and live in Hamden, Connecticut.